Improving Irrigation in Asia

Improving Irrigation in Asia

Sustainable Performance of an Innovative Intervention in Nepal

Elinor Ostrom

Senior Research Director, Workshop in Political Theory and Policy Analysis, Professor of Political Science, Indiana University, USA and 2009 Winner of the Nobel Prize in Economic Sciences for research on economic governance

Wai Fung Lam

Professor of Politics and Public Administration, The University of Hong Kong

Prachanda Pradhan

Patron, Farmer Managed Irrigation Systems (FMIS) Promotion Trust, Nepal

Ganesh P. Shivakoti

Professor of Agricultural and Natural Resource Economics, Asian Institute of Technology, Thailand

Edward Elgar

Cheltenham, UK • Northampton, MA, USA

Published by
Edward Elgar Publishing Limited
The Lypiatts
15 Lansdown Road
Cheltenham
Glos GL50 2JA
UK

Edward Elgar Publishing, Inc.
William Pratt House
9 Dewey Court
Northampton
Massachusetts 01060
USA

A catalogue record for this book
is available from the British Library

Library of Congress Control Number: 2011927319

ISBN 978 1 84980 144 7 (cased)

Typeset by Servis Filmsetting Ltd, Stockport, Cheshire
Printed and bound by MPG Books Group, UK

Contents

Tables

Abbreviations

ADB	Asian Development Bank
ADB/N	Agricultural Development Bank of Nepal
AKRSP	Aga Khan Rural Support Programme
AMIS	Agency-managed irrigation systems
CARE	Cooperative for American Relief Everywhere
CSIP	Community Surface Irrigation Project
DAO	District Administrative Office
DDC	District Development Committee
DIO	District Irrigation Office
DOI	Department of Irrigation
DOLIDAR	Department of Local Infrastructure Development and Agricultural Roads
FAO	Food and Agriculture Organization of the United Nations
FIWUD	Farm Irrigation and Water Utilization Division
FMIS	Farmer-managed irrigation systems
HFPP	Hill Food Production Project
HMG/N	His Majesty's Government of Nepal
IAAS	Institute of Agriculture and Animal Science
IDD	Irrigation Development Division
IDSD	Irrigation Development Subdivision
IIMI	International Irrigation Management Institute
ILC	Irrigation Line of Credit
ILO	International Labour Organization
IMSSG	Irrigation Management Systems Study Group
IMT	Irrigation management transfer
ISP	Irrigation Sector Project
ISSP	Irrigation Sector Support Project
IWMI	International Water Management Institute
MLD	Ministry of Local Development
NIIS	Nepal Irrigation Institutions and Systems
NRs	Nepalese rupees

QCA	Qualitative Comparative Analysis
REDP	Rural Energy Development Project
SES	Social-ecological system
SINKALAMA	Sindhupalchok, Kavreerpalanchok, Lalitpur and Makwanpur
SISP	Second Irrigation Sector Project
SPA	Seven Party Alliance
UMN	United Mission to Nepal
UNDP	United Nations Development Programme
USAID	United States Agency for International Development
VDC	Village Development Committee
WECS	Water and Energy Commission Secretariat
WUA	Water User Association
WUC	Water User Committee

Foreword

Gilbert Levine

Investment in irrigation in developing countries by the major lenders during the past half-century has generally been more economically successful than comparable investment in other agricultural programs. Nevertheless, many of the efforts in the past three decades have not achieved the desired results, and at least some would be considered failures. Why, with all of the professional expertise, with all of the money invested, and with major needs, has this occurred?

I suggest that there is a two-part answer. The first is exemplified by this quote from H.L. Mencken: 'For every complex problem there is a simple solution . . . and it is wrong.' The second is the failure to believe in the concept of *necessary fallibility*.

Realization of the very complex nature of the problems associated with irrigation system design and improvement is relatively recent, though recognition that more needs to be considered than the physical works and the training of operators is not new. Both research and experience have suggested that greater attention should be given to the non-physical aspects of water capture, delivery, and use, but the internal objectives and constraints of major donors and related national agencies have often limited such consideration. In place of designs and implementation tailored to individual situations, models that *simplify* the complex problems have been used. The Taiwan model of irrigation association, with rotational irrigation as a centerpoint, was prevalent in the 1980s with limited success. More recently, the 'Mexico mode' of water user organization has been the basis, often with a token emphasis on the experience of farmer-managed irrigation systems. The results have frequently been less than anticipated. Even the application of these models has been problematic because those involved in planning and implementing improvements were often constrained in their ability to carry them out. Terms of reference, whether by the international lenders and donors, or by the responsible national offices, not infrequently imposed time limits for planning and implementation that prevented appropriate interaction with the end users; unrealistic requirements for user participation resulted in _nominal_ rather than _actual_ participation; conflicts of interest, especially when the agencies and individuals involved may be adversely affected by the results, have delayed and prevented implementation. However, even as understanding of the need to address the problem of the complexity of the human-irrigation environment has grown, the acceptance of our inability to fully do so has not.

This brings me to necessary fallibility. In many areas of human endeavor, there is either insufficient information, or understanding, or available resources, individually or in concert to bring about successful action. Nowhere is this more prevalent than in the area of irrigation system design and improvement where there are multiple smallholder users. The cost of obtaining the necessary information of the physical environment as well as the social environment at the scale required for successful design would be prohibitive, even if we knew what information was necessary and the methodologies for obtaining it were available. Thus, under these constraints, there must be recognition that the initial efforts _necessarily_ will have mistakes and inadequacies. The first implementation, unavoidably, will be an 'approximation.' To permit appropriate modification in a 'second approximation,' sufficient monitoring, feedback, and resources should be included to identify and to implement corrections and changes.

As the irrigation community moves from the era of oversimplification to acceptance of the complexity inherent in the sector, research such as that presented in this book provides understanding that, at least, will help to make the 'first approximation' a better one. At best, it provides an approach that addresses both the complexity and fallibility problems.

Gilbert Levine
Professor Emeritus
Mario Einaudi Center for International Studies
Cornell University
Ithaca, NY

Foreword

Robert Yoder

Reviewing the chapters of this book has allowed me to revisit the activities that brought me into contact with the authors. I find great satisfaction in seeing that action-research initiated in Nepal to explore innovative interventions in farmer-managed irrigation systems (FMIS) in the mid-1980s has received sustained attention for over two decades. Imperfect as action-research often is in setting a solid baseline for time series analysis, investigation subsequent to the completion of the 'project' has added much to the lessons gleaned from the 'action' part of the research. The 'action' effort was to bring emerging information about FMIS into activities endeavoring to improve water use and expand the area irrigated. The focus was to enable the irrigation users to gain formal recognition as Water User Associations (WUAs) and to stay in control of system operation and maintenance as well as the planning and implementation of future initiatives.

Studies of FMIS in Nepal in the 1970s (Bihari Krishna Shrestha) and early 1980s (Prachanda Pradhan, Edward Martin, and Robert Yoder) spawned field investigation of FMIS by numerous additional scholars over the next decades. Some of Nepal's FMIS are hundreds of years old and were built by feudal landholders, religious trusts, and the local community of the farmers. A tradition of local governance was established where water management was the responsibility of the users with no interference by the king's court or by the state governing machinery. The early success of locally managed irrigation prompted farmers all over Nepal, as in many other countries around the world, to organize collective action to build, operate, and maintain additional systems. Decade by decade, new diversions and canals were constructed so that by the end of the nineteenth century, FMIS had spontaneously evolved in most areas where both water and land are accessible in Nepal.

The early research provided insights that exposed unheralded resources. Foremost among these was that multitudes of locally built and governed stream diversion irrigation systems are successfully irrigating many thousands of hectares of crops in both the hill and *terai* areas of Nepal. Another finding was that cropping systems have evolved from one crop to two and three crops per year depending upon altitude. Also, that the food produced by these irrigation systems is essential in supporting a significant percentage of Nepal's population.

In addition to the technical sophistication of aligning canals that wind around mountains and cross sheer rock cliffs, sometimes threaded through tunnels – almost exclusively built by local labor and tools – it was determined that FMIS success is based upon indigenous governance systems that evolved as the systems were planned and constructed. The studies highlighted the ability of water users to devise, continually adjust and enforce the roles and rules by which their systems are operated and maintained. They concluded that the ability to craft institutions that fit local conditions is a valuable and underutilized resource. Capturing this largely invisible resource as a driver for irrigation development became the challenge that emerged from the FMIS field studies.

The Water and Energy Commission Secretariat (WECS) and newly renamed International Water Management Institute (IWMI) determined to use information from the field studies to investigate ways of addressing the escalating cost and disappointing performance of hardware-based irrigation development. With support from the Ford Foundation, they developed an action-research activity to look at ways to incorporate FMIS resources into Nepal's push to expand irrigated agriculture. The action-research was an experiment in using physical construction activities to strengthen self-governing and self-supporting irrigation institutions at the user-community level.

The authors of this book looked at the action-research project results in the late 1980s and incorporated the project irrigation systems into their ongoing global research on local resource management. They found ways to revisit and review what happened to the systems over the 20-year period following the WECS/IIMI intervention. This book provides an excellent summary of the action-research experiment but goes much further by reporting on observation and analysis of the project impact spanning the two-decade, post-project period. This adds a new dimension to 'action-research' and gives information about the dynamics, benefits, and limitations of local governance. The insights obtained from the longitudinal study provide a guide for refining the process of supporting the growth and expansion of FMIS and input to be used in updating intervention strategies for all types of community-based resource management.

Robert Yoder
Technical Director Water Technology
International Development Enterprises
Addis Ababa, Ethiopia

Acknowledgments

We wish to acknowledge the many organizations and individuals who have directly or indirectly contributed to the completion of this volume. The study has evolved from two impetuses: the works and data of the Nepal Irrigation Institutions and Systems (NIIS) project at the Workshop in Political Theory and Policy Analysis (the Workshop) at Indiana University; and the then International Irrigation Management Institute (IIMI) Nepal country office, which was a major source of case studies that were used for coding the data sets for the NIIS project. A subset of data sets provided by IIMI is the major source of analytical book chapters contained in this volume. Matrika Bhattarai, Ujjwal Pradhan, Robert Yoder, and Naresh Pradhan were instrumental in providing all the necessary support at both the intellectual and field study levels. We would also like to extend our sincere thanks to the farmers, irrigation officials, and several local organizations in the Indrawati watershed basin of Sindhupalchok District of Nepal for their hospitality and help.

We have benefited tremendously from the individual and organizational support provided by Walter Coward of the Ford Foundation in New York and John Ambler of the Ford Foundation in New Delhi. We wish to acknowledge the institutional support and permission to use the publications provided by the Water and Energy Commission Secretariat (WECS), Nepal, and International Water Management Institute (IWMI), Sri Lanka. The research was also supported by the National Science Foundation, Ford Foundation and the Workshop.

Colleagues at the Irrigation Management Systems Study Group (IMSSG) of the Institute of Agriculture and Animal Science in Rampur, Nepal – Ashutosh Shukla, Tej Bahadur KC, Neeraj Joshi, Narayan Joshi, Kala N. Pandit, Kailash N. Pyakuryal, Shanta Man Shakya, Deo Nath Yadav, and Kishor Gajurel – were involved in collecting field data during different phases of data collection. Similarly, colleagues at the Workshop – Julie England, Paul

Benjamin, Gopendra Bhattrai, Sharon Huckfeldt, and Myungsuk Lee – have contributed a substantial amount of time and effort in refining concepts, thinking through theoretical questions and processing data.

The authors also thank Marty Anderies, Eduardo Araral, Xavier Basurto, Sue Crawford, Marco Janssen, Edella Schlager, Pontus Strimling, and Rick Wilson for comments on earlier versions of Chapter 4, and Charles Ragin for offering helpful suggestions for the QCA analysis in the study. We wish to acknowledge the analysis and write-up support provided by a grant from the Hong Kong Research Grants Council (HKU7233/03H) to the University of Hong Kong and by the National Science Foundation (Grant BCS–0527744) to Arizona State University and Indiana University. The several rounds of meetings among authors from the Asian Institute of Technology (AIT), FMIS Promotion Trust of Nepal, the University of Hong Kong (HKU), and the Workshop were made possible through a Seed Fund (200911159155) from HKU and a grant from the country office of the Ford Foundation in Delhi to AIT, which are duly acknowledged.

Permission is granted by Water and Energy Commission Secretariat, Nepal, and International Water Management Institute, for Tables 3.2 and 3.3 in this volume from *Assistance to Farmer-Managed Irrigation Systems: Results, Lessons, and Recommendations from an Action-Research Project* (Colombo, Sri Lanka: Water and Energy Commission Secretariat, Nepal, and International Irrigation Management Institute, 1990, pp. 29–30). There are partial sections in Chapter 4 of this volume that draw on an earlier article by Wai Fung Lam and Elinor Ostrom (2010), 'Analyzing the dynamic complexity of development interventions: Lessons from an irrigation experiment in Nepal', *Policy Sciences*, **43** (1): 1–25. Tables reprinted with kind permission from Springer Science+Business Media B.V.

We appreciate the excellent contributions of Gilbert Levine and Robert Yoder in writing the Forewords for our book. This book would never have been completed without the endless editing efforts of Patty Lezotte to put everything together in a coherent manner.

1 The challenge of achieving successful development interventions

THE PUZZLE: HOW TO HELP FARMER-MANAGED IRRIGATION SYSTEMS

According to the Food and Agriculture Organization (FAO) projections, the world's food demand between 2000 and 2030 is expected to increase by 55 percent (FAO, 2003). To meet the increased demand, world food production has to increase at an annual rate of 1.4 percent (UNDP, 2006). Much of this growth would need to occur in developing countries. Given the limited potential for increasing the amount of cultivable land, much of the increased food demand would have to be met through improved agricultural productivity and increased cropping intensity (UNESCO-WWAP, 2006; Molden, 2007). Improving irrigation systems' performance is thus a key production factor that could help lift the constraint on agricultural productivity in developing countries in Asia and elsewhere. In fact, cross-country comparisons have found that irrigation is strongly associated with the level of the risk of poverty and economic inequality (UNDP, 2006). It has been a high priority for the allocation of international aid to generate and maintain irrigation robustness in Asian countries.

An irrigation system is robust if it is able to keep water at target levels within the system under uncertain environmental disturbances for long periods of time. The engineering aspect of an irrigation system can be threatened by two disturbances coming from the broader environment. One is the threat of flooding due to excessive rainfall. The second is the result of too little rainfall and the consequent inadequate supply of water for the farmers. The way engineering works try to protect against flooding is the construction of very strong weirs, gates, canals, and other control mechanisms that enable the system to survive even though water is pouring down from the hillsides and stressing the strength of the control mechanisms and other engineering parts of the system. The way that irrigation systems try to protect against drought is the provision of storage in the system that holds water from wet periods to be used in dry periods. Many run-of-the-river systems, however, have no facilities for storage. Thus, another major mechanism for coping with drought has to do with the institutional rules used by a system for allocating water under systems of scarcity (Carruthers, 1981; Burns, 1993).

Institutional rules are also important in regard to the maintenance of the engineering works themselves. All irrigation diversion structures and canals, whether constructed from rock, mud, or concrete, need regular maintenance. The maintenance must be done either by the farmers sharing the workload, by officials of an irrigation system for an agency-managed irrigation system, or by both when the system is co-managed (Lam, 1996a, 1998, 2006b). The crucial concern is that there are either cooperative arrangements developed for farmers to contribute their labor or an effective taxing system that contributes the necessary resources to provide the essential resources for maintaining a system.

Prior studies have provided ample evidence that farmers in some irrigation systems have evolved rules over a very long period of time at multiple levels that enable their systems to be robust (Barker et al., 1984; Ostrom, 1990; P. Pradhan, 1989a, 1989b). Even systems that have been robust for multiple decades or even centuries, however, may face additional threats beyond flooding and drought. If external authorities do not understand the way a system is organized, they may make top-down changes that threaten the foundation of the institutional arrangements that keep such systems going over the long run (Lam, 2005; Ostrom, 2005; Vermillion, 2005).

A puzzle that has presented itself to many governments and donor agencies is what to do when farmer-constructed and operated irrigation systems do not appear to be robust given several indicators of stress or poor performance. One of the most important measures is when farmers are not able to grow a substantial crop due to the unpredictability or inadequacy of water supply at key times in the agricultural cycle. Another indicator is the loss of regular support from farmers for contributing time and effort to rebuild key parts of the system or for regular maintenance (Araral, 2009). Systems may also lose their robustness when economic conditions change so that occupations other than farming become more attractive and lure younger families away from the region (Lam, 2001; Baker, 2005; Shivakoti et al., 2005). Obviously, conflict in a region is also a major disturbance that may disrupt how irrigation systems and other core activities operate.

When governments or international donors observe faltering irrigation systems, they sometimes simply come in and build entirely new systems. While that provides new engineering works, it has frequently led only to temporary improvements in the performance of the irrigation system (Chambers, 1988; Johnson, 1991; Ostrom et al., 1993; Lam 1996b). Further, systems that have received major external help in reconstruction sometimes become totally dependent on external help. But, what is to be done? How can an external agency help irrigation systems that are faltering without leading them to become dependent on external aid?

INTERVENTION AS A SEARCH FOR 'BEST PRACTICES'

Over the past decades, international donors have allocated substantial funds to adopt the 'best practices' in an effort to improve the performance and robustness of irrigation systems. Among the best practices that have been introduced and tried out, two stand out as the most influential and have provided the rationale upon which intervention policy is formulated and implemented. The first is to hire external water engineers to construct 'modern' irrigation engineering infrastructure to replace the primitive structure in use by farmers; the second is to put in place institutional templates for the organization of collective action for irrigation management.

Infrastructure investments have frequently led to temporary improvements, but not longer-term enhanced performance (Chambers, 1988; Lam, 1996b). Likewise, the introduction of 'institutional reforms' in systems has often brought about only institutional change in many instances. In some instances, the user groups that the institutional reforms sought to build existed only in name. Experts who have examined the performance of the systems that were allocated development assistance have rarely given positive reviews (Yudelman, 1985). Hugh Turral, for example, concludes that 'irrigation schemes have often underperformed in economic terms, and field research has highlighted substantial shortcomings in management (operation and maintenance), equity, cost-recovery, and agricultural productivity' (1995, p. 1). Further, systems that have received major external help then become totally dependent on external help (Araral, 2005).

Comparing Farmer-managed and Agency-managed Systems in Asia

The first writings examining the processes of intervention focused on comparing farmer-managed irrigation systems (FMIS) and agency-managed irrigation systems (AMIS) in the Philippines (de los Reyes, 1980; Siy, 1982; de los Reyes and Jopillo, 1988; Sengupta, 1991). Subsequently, a major effort to document public intervention in FMIS in the Asian region was made by the International Irrigation Management Institute (IIMI) during an international conference in Nepal in 1986 (IIMI/WECS, 1987). A common and important theme of the research findings from the conference was that '[b]efore intervening, agencies should understand how the existing farmer-managed systems are organized, the way they carry out irrigation activities, and the environment in which they operate' (Martin and Yoder, 1987, p. iv).

In Sri Lanka, Medagama (1987) found that the Department of Agrarian Services' Village Irrigation Rehabilitation Program suffered from a lack of information as well as farmer participation in the planning process, which inhibited the rehabilitation process of small-scale village tank and diversion systems and the initiation of the water management program. Moreover, since the rehabilitation process and water management were viewed separately and were undertaken by two different departments, it reduced the capacity of the farmer organization to manage the system, which resulted in greater dependence on the government.

An evaluation of the irrigation component of the Aga Khan Rural Support Programme (AKRSP) in Pakistan, on the other hand, pointed out that the program has helped to build effective local institutions that can select and implement development programs (Hussein et al., 1987). The village organization was one such local institution where members of the organization chose the kind of activity they wanted to be financed and were involved at all stages of the program implementation. According to the authors, the policy of paying for local labor, which counted as part of the grant, as well as disbursing grants in installments, were important features of the intervention that resulted in an irrigation program that was technically feasible, institutionally sustainable and economically profitable.

In the Philippines, Ben Bagadion (1987) described how academic research on FMIS influenced the Philippines' (National Irrigation Administration) approach to test and modify the intervention process through action-research. The important lessons learned from action-research on FMIS interventions pointed to the necessity of farmers' participation in planning, construction, and organizational development, as well as having all stakeholders work together through a learning process approach (Korten and Siy, 1987). Similarly, action-research conducted in northern Thailand confirmed the importance of increasing interaction, communication, and coordination among agency staff, farmer irrigators, and researchers (Tan-Kim-Yong, 1987). Furthermore, the research argued for using more farmer-to-farmer training and consulting services that are backed up by a mobile team of professionals.

Further, one paper examined why agencies in Indonesia rarely utilized the results of the research by contrasting the macro- and micro-policy arenas as well as policy analysis and social learning perspectives found within the literature (Korten, 1987). Korten argued that most issues related to policy intervention in FMIS fell into a micro-policy arena and, therefore, the social learning research is more appropriate for addressing these issues. Another aspect of the intervention issue is how to hand an entire system back to the farmers without making them feel the burden of the additional and often expensive operation and management costs.

The Best Practice of Investing in Better Engineering Works

Decades of efforts to help farmers improve irrigation performance in developing countries in Asia have been underlined by the assumption that assistance is primarily concerned with, and hence hinged upon, an improvement of irrigation engineering infrastructure. To many officials in charge of assistance programs in governments and international donors, particularly those with an engineering background, irrigation is about delivering adequate amounts of water to the right place at the right time. Thus, to enhance the delivery capacity requires that strong engineering works be built for better control of the volume and flow of water (Wade, 1987; Chambers, 1988; Uphoff et al., 1991; Lam, 1998). Naturally, professional engineers who command the knowledge of hydraulic and civil engineering are expected to play a leadership role in providing technological blueprints that help maximize water delivery efficiency.

Such an engineering-centered approach to irrigation assistance very much resonates, and in fact might have owed its origin to, two ideas that have been dominant in the development communities – the idea of a 'poverty trap' being the major cause of underdevelopment and the idea of a 'big push' being the solution for helping the poor to get out of the poverty trap (Easterly, 2001, 2006; Rodrik, 2007). The crux of the 'poverty trap' argument is that people in developing countries are poor because they started poor; the lack of endowments and opportunities has prevented the poor from reaching a threshold level of capacity necessary for sustained development, and hence created a vicious circle of poverty. Once trapped, the poor will hopelessly stay at the bottom and can never get out of the trap by themselves.

To help the poor break out of the poverty trap, a big push of external assistance is needed that presumably will help them pass the threshold so as to jump-start the spiral of development (Bowles et al., 2006). The engineering-centered approach to irrigation assistance seems to be based on the same big-push logic. The argument is that many irrigation systems in developing countries in Asia have failed to perform because they do not have what it takes to control and deliver water effectively. By building stronger and more sophisticated engineering works, farmers in the system are given the necessary capacity to do a better job in controlling water in the canals. An implicit assumption is that once effective control of water is put in place, good irrigation performance will follow.

A caveat is warranted here. We would not do justice to irrigation policymakers and managers, particularly the irrigation engineers, if we argued that all of them were ignorant of, or had turned a blind eye to, the fact that irrigation management is, by and large, a social-technological process. In fact, with years of intervention experiences and advocacy by social scientists, policymakers and specialists in the irrigation policy community in Asia are in general aware of the importance of the social-institutional dimension of irrigation management. Such an awareness is reflected by the fact that, with few exceptions, assistance projects nowadays have included 'institutional building' as a major project component. But recognizing the importance of the social-institutional aspect of irrigation management is one thing; appreciating that an irrigation system is a social-ecological system (SES) involving complex interactions between human actions and physical-biological dynamics is quite another matter. Very often, the construction of engineering works and institutional development are considered to belong to two related yet separate domains. The usual assumption is that only after improved engineering works have been put in place could 'appropriate' institutions be molded in accordance with some best-practices templates for the operation and maintenance of the infrastructure.

The major inadequacy of the engineering-centered approach to irrigation assistance is not that engineering works are not important; of course they are. But engineering works are but one of many components that constitute an SES of which an irrigation system is a prototype. Janssen and colleagues define an SES as:

> composed of biophysical and social components where individuals have self-consciously invested time and effort in institutional infrastructure (and, in some cases, physical infrastructure) that affects the pattern of outcomes (e.g., patterns of resource use and their distribution within the population) achieved over time in coping with diverse external disturbances and internal problems. (Janssen et al., 2007, p. 309)

An irrigation system composed of a resource (sources of water), physical infrastructure (storage and canals), actors who manage and appropriate from the resource (farmers and irrigation managers), and a governance structure that regulates the action and interaction of the actors (irrigation institutions) is an example of an SES. An SES is a complex system. Its features emerge from the interactions of actors within the system. Its dynamics are activated by human and biophysical processes at multiple spatial and temporal scales and scopes that often generate complex positive feedback loops. It is often nested within SESs of larger scales and scopes (Miller and Page, 2007; Mitchell, 2009; Ostrom, 2009).

Given the complexity involved in an SES, attempts to intervene or 'assist' the system pertain to harnessing the operation and dynamics of the SES with a view to improving its performance and robustness (Axelrod and Cohen, 2000; Carlson and Doyle, 2002). An effective intervention process has to be designed in conjunction with the operation of the SES rather than being conducted as an external process of manufacturing changes to the system. In Nepal and many other Asian countries where the engineering-centered approach has been, and in fact is still very much, dominating the assistance community, an intervention project is often considered to be no more than a package of 'deliverables' to be provided by government or donor agencies. Officials in these organizations often look at an intervention project through a bureaucratic lens, focusing on how to manufacture the deliverables in accordance with some criteria and standards specified by their organizations. A result is that the process of intervention has often been turned into a sequence of bureaucratic decisions oblivious of the actual operation of the irrigation system (an SES) that the intervention is supposed to help. When the question of how assistance resources could be used to help the system users becomes one of budgetary allocation, officials look at the process of intervention as no more than a checklist of administrative procedures. It is unlikely that the 'deliverables' would fit what is needed for improving the performance of the SES.

Given the complexity and uniqueness of an SES, any attempt to 'intervene' in the system is inevitably an ongoing concern. It requires continuous inputs from individuals involved who have first-hand experience of how the intervention operates, and also feedback mechanisms that allow continuous adjustments and adaptations in the design of the intervention programs to the changing environment (Berkes and Folke, 1998; Berkes, 2002; Lam, 2006a). An engineering-centered approach to irrigation assistance, unfortunately, is by its very nature antithetic to an emphasis on user feedback and continual adjustments. Given that the construction of engineering works is presumably based on technical knowledge, it is self-evident to many policymakers that professional engineers who command the technical knowledge are in the best position to tell the system users what kind of engineering works they need or should have. A consequence is that engineers and farmers find themselves in an asymmetric power relationship in which the former see themselves as the 'help providers' and the latter the 'help recipients.' In such an asymmetric relationship, one would not be surprised that the voice of the 'help recipients' is not taken seriously by the 'help providers.'

The power asymmetry not only hinders feedbacks but, more importantly, distorts the accountability relationships of officials working in donor or government agencies vis-à-vis users of the system that is being assisted (Lam, 1998; Easterly, 2006). Making a person accountable is to make him or her answerable for his or her decisions or actions and, more importantly, bear the consequences. Accountability can be construed with reference to different criteria or standards, with varying scopes; and the way that accountability is construed can have serious implications for the design and implementation of an intervention project. In the engineering-centered approach, officials in charge of the implementation of intervention are often made accountable to their superiors in their respective organizations; whether and how the officials' decisions or actions impact on the operation and performance of the irrigation system being assisted often becomes of marginal significance to them. Such an incentive structure encourages the officials to focus more on satisfying their superiors and making sure the list of deliverables are checked off, than on solving problems for system users.

The issue of getting the accountability right is in fact not confined to the relationship between officials and system users but also among the users themselves. When system users perceive that an intervention project is a game in which they compete with one another for benefits from the outside, the project destroys rather than strengthens the social capital of local communities (Gibson et al., 2005; Shivakumar, 2005; P. Pradhan, 2010). To make both system users and officials accountable for the collective good of effective intervention, it is important that the intervention process be designed and implemented in a transparent manner. Transparency not only facilitates monitoring but, more importantly, helps generate and sustain a common cause.

Another pitfall of an engineering-centered approach is that the process of intervention tends to focus on maximizing technical efficiency as the single most important performance measure. Scholars who are deeply concerned with the complexity involved in an SES have long argued for the importance of paying due attention to the implications of an efficiency–robustness trade-off for the operation of an SES and, hence, for the design of intervention efforts seeking to improve the performance of the system (Anderies et al., 2004; Janssen et al., 2007). Engineers are trained to build infrastructure that maximizes water control capacity in relation to specific hydrological and physical constraints. The more the engineering works are aligned to the constraints, however, the more likely that the works become so highly optimized to the specific constraints that the system becomes extremely sensitive to other types of external disturbances such as landslides or drastic social-economic changes in the region; technical efficiency is often attained at the expense of the system's robustness. The efficiency–robustness trade-off might explain why many assistance efforts in the past tended to have rather a short-lived impact on performance, and failed to sustain in the long run.

The Best Practice of Searching for a Good Institutional Template

Another best practice of irrigation assistance in the past decades is concerned with a search for a good institutional template for managing the operation and maintenance of irrigation systems (Lam, 2011). In fact, the development and evolution of irrigation assistance policy in Asia in the past decades has been underlined by a search for the best institutional arrangements for irrigation management. Until the 1970s, the policy recipe for effective irrigation management was to put governmental authorities in place to play the roles of a provider, a regulator, and a promoter of irrigation development (Coward, 1980; Barker et al., 1984). Based upon the policy recipe, international donors were advised of the best practice of channeling adequate monetary aid to help construct large-scale, sophisticated irrigation infrastructure and to replace primitive systems in use (Easter, 1986; Ostrom et al., 1993). Also, effective operation of the sophisticated irrigation infrastructure is hinged upon 'professional' irrigation management, which in practice was often translated into the setting up of a centralized irrigation bureaucracy and the development of sophisticated technical guidelines for irrigation operation and maintenance (Chambers, 1988; Ascher and Healy, 1990; Lam, 1996b). Many Asian countries even went to the extreme of implementing a large-scale nationalization of irrigation systems.

By the early 1980s, there was strong evidence that the bureaucratic mode of governance was not up to the seemingly straightforward, simple task of moving water to the right place at the right time. A study of more than 150 irrigation systems in Nepal, for example, found that systems that were characterized by sophisticated irrigation infrastructure and professional management by engineers in government agencies were outperformed by systems in which the dams and canals were built with primitive materials and the operation and maintenance were performed by farmers in the systems (Lam et al., 1997; Lam, 1998). Such counterintuitive outcomes were not confined to Nepal but were found repeatedly in many other Asian developing countries.

The disappointing performance of the bureaucratic mode of management triggered a search in both the academic and policy communities for reasons for the failure and also possible alternatives. There are two often-cited diagnoses. First, irrigation infrastructure investments were often made based on the criteria of technological efficiency and effective demand for irrigation water. The simple fact that irrigation involves more than the mechanical operation of irrigation infrastructure, but also a productive working order among systems' managers and users, has often been ignored. Second, the bureaucratic mode of governance has not paid due attention to the motivation of the irrigation officials in irrigation management, nor to the incentives faced by the farmers using the systems (Wade, 1982; Chambers, 1988; Tang, 1992; Lam, 2006b).

As a response to the disappointing performance of the bureaucratic mode of management, many policymakers turned to markets for alternative solutions. Most of the applications of the market approach have emphasized privatizing the rights of appropriation from the resources. A market-oriented governance structure usually combines governmental ownership with a system of exchange of the rights of appropriation (Dinar et al., 1997). The typical institutional arrangement is that governmental authorities own the water resource and determine both the aggregate level of appropriation allowed and the allocation of the rights of appropriation on the basis of some macro-policy criteria. An individual user who is allocated water rights can exercise the rights for their own use or trade them in a water market. The governmental authorities will play the role of an arbitrator to enforce the water rights and to oversee the operation of the water market.

While market-oriented governance seems to provide a theoretically sound recipe for coping with the collective-action problems involved in irrigation management, practical experiences suggest that introducing a market mode of governance is a complicated process involving negotiation within wider contexts and preconditions such as political choices and cultural practices (Bauer, 1997; Whitford and Clark, 2007). Also, effective operation of water markets requires that the transaction costs involved in the exchange of rights be kept at a reasonably low level. In particular, how to maintain a clear common understanding among owner-users about the proper functioning of the water market is a big challenge.

Irrigation research and assistance experiences alike have found more complicated pictures that defied simple association between a particular policy panacea and irrigation performance. For instance, the failure of the bureaucratic mode of irrigation governance has built in substantial skepticism in many policymakers who believe that government is inherently inconsistent with effective irrigation management. Studies of social capital and development, on the other hand, suggest that a state–society synergy is instrumental to materializing development potentials in various domains of collective action (Evans, 1996; Lam, 1996a). In irrigation management, research has also suggested that synergy between farmers and government officials holds the key to irrigation performance (Wade, 1982; Moore, 1989; Lam, 1996a, 2006a, 2006b). Consequently, the study of the governance of irrigation should not focus on a choice between markets and the state. Instead, more attention should be given to the kinds of working relationships among resource managers and users that are more likely to bring about good performance and the kinds of governing mechanisms needed to sustain the productive relationships (Curtis, 1991).

Evidence from many communities of resource users around the world has suggested that resource users not only cope with collective-action problems but they are also capable of crafting rules to govern their own systems. The potential of self-governance is by no means trivial. In irrigation, for example, some of the oldest, most long-enduring systems have been constructed, governed, and maintained by farmers themselves for long periods of time (Cernea, 1987; Shivakoti et al., 2005). Research also suggests that, even in irrigation systems that are governed and managed under a bureaucratic mode, a certain degree of self-governance among farmers at the field level is essential to effective operation of the systems (Chambers, 1988; Wade and Seckler, 1990; Uphoff et al., 1991; Lam, 1996a).

The (re-)discovery of the potential of farmers' participation in irrigation management triggered a wave of reform efforts to organize local farmers and to enhance their involvement in irrigation management. Research on common-pool resources and, in particular, irrigation management, has found that self-governing institutions that allow farmers to design their own rules to govern themselves are more likely to bring about productive working relationships, and hence better irrigation performance. Building upon serious theoretical analysis and extensive fieldwork, scholars have identified aspects of institutional design that are conducive to self-governance.

In many developing countries in Asia in the last two decades, building self-governing capacities in communities of farmers through such measures as setting up a Water User Association (WUA) and management transfer programs has become a component in many assistance projects funded by national governments and international donors. In countries where there is a dominant public irrigation sector, pilot projects to transfer management from government to farmers in irrigation systems were tried out. While there are successful examples, there are more exceptions than norms.

The crux of an irrigation management transfer (IMT) initiative is a rethinking of the division of labor, power, and responsibility between irrigation officials and local farmers (Bruns and Atmanto, 1992; Vermillion, 1997). It has been generally recognized that farmers and irrigation officials possess respective information and resources that the others do not have. Hence, neither one of them can effectively manage irrigation acting alone. The question of exactly what the division of power and responsibility should look like in a new order of work, however, is largely a guessing game. As a result, in some instances IMT was implemented as an exercise in which irrigation officials 'advise' the farmers what to do and how to supplement their effort in the plans that were laid down by the officials. In some other cases, IMT was conducted as a retreat of government resources and involvement from irrigation.

Building self-governance and enhancing farmers' participation can take another form for systems that are managed by farmers. Many international donors have made 'institutional development' a required component of their assistance schemes. With the help of an army of agricultural extensionists and community organizers, farmers were encouraged to organize into WUAs to manage their own systems. In fact, being able to organize a WUA has often been made a precondition for farmers in a system to receiving assistance from international donors. Similar to the experience of irrigation management transfers, the efforts to jump-start WUAs have had only mixed results.

If there was one lesson that decades of efforts to 'develop' local institutions had to offer, it would be that institutions cannot be imposed or transplanted easily as if they are physical capital. International donors can certainly arrive at a system and proclaim well-intentioned rules that were thought to be good for the farmers and more importantly self-sustaining and self-enforcing. But if the farmers do not understand or take these rules seriously, the well-intentioned attempts to put self-governance in place have brought about nothing but facades of rules-in-form that, in many cases, farmers were not even aware of. While 'building local self-governing institutions' seems to be a simple, straightforward panacea to some policymakers, putting in place institutions that work has proven to be a substantial challenge. We may know what general kinds of institutions are more conducive to self-governance, but we do not know how to help bring about these institutions.

Decades of search for the best institutional template have come back to ground zero, with a serious puzzle. What is puzzling is that policymakers and policy analysts now tend to agree that institutions matter. Their understanding of these institutions, however, has not gone very far beyond looking at them as static templates of institutional design, a blueprint on paper. While we might have good ideas about what 'good' institutions, and particularly self-governing institutions, look like, we have rather limited knowledge as to how these institutions actually operate, adapt, and evolve over time; much less do we know about how to harness these institutions so as to build, enhance, and steer the process of self-governance, if that is at all possible.

LESSONS FROM INTERVENTION EXPERIENCES

Although the decades of intervention efforts have provided only mixed results, they have provided valuable experiences for a critical reflection of what intervention is about and how we could approach it in a more fruitful manner. In particular, several lessons can be learned from the experiences.

First, the Holy Grail of a unique universal set of rules that can bring about efficiency and robustness in irrigation does not exist. Policy analysis for assistance should move away from the elusive quest for best practices and give due recognition to the complex sets of factors and processes that constitute the problems in a particular situation. All policy panaceas are necessarily based upon an often simplified notion of undesirable phenomena involved; possible undesirable phenomena in irrigation management include poor infrastructure, lazy and malicious farmers, unclear property rights, unorganized farmers and so on. With the undesirability identified, whatever it might be, it is presumed easy to infer a set of 'desirable' conditions or behavior of individuals to be translated into 'the' most appropriate institutional arrangements.

The attractiveness of policy panaceas is that they simplify by telling the policymakers what is wrong and what solutions they should adopt to correct the wrong. Unfortunately, all irrigation systems are not alike; each of them has unique features and characteristics as they operate in different sociocultural and environmental contexts. Irrigation management, like many other collective human endeavors involving complex social-ecological interactions, cannot be easily simplified without a cost. Specifically, a particular undesirable phenomenon could be caused by a whole array of processes, depending on the social-ecological context of an irrigation system. Poor infrastructure, for example, could be a result of inadequate resources. Yet, it could also be a measure adopted by local farmers to cope with some ecological features of their systems. Prachanda Pradhan (1989b), for example, reports that for many systems in the Tanahu District, Nepal, the temporary nature of dams or intakes is in fact part of the mechanism to control water flow in the systems. Building a permanent intake structure in these systems would only reduce their robustness to the fluctuation in water flow at the source of the system. By the same token, a particular (desirable) behavior pattern of farmers could possibly be attained by different institutional arrangements, depending again on various social-ecological factors and processes in different contexts. Assuming a linear mapping between a particular outcome and a particular institutional arrangement not only risks what Peter Evans (2004) calls institutional monocropping but also unnecessarily limits the scope of possibilities.

Second, contexts matter. Irrigation management involves a large array of social-ecological factors and processes that operate and interact in a complex manner. Very often it is the combinatorial effects of several factors that generate a particular problem or outcome; and the combinatorial effects are usually related in a nonlinear manner. A common problem in different contexts could require very different solutions and hence institutional arrangements. Once one recognizes that complexity is unavoidable, it becomes obvious that a search for policy panaceas, which is largely based on a linear thinking mode, is an impossible endeavor. An important implication for policy analysis is that, instead of trying to isolate a small number of variables and to examine how they might affect outcomes, an analyst has to adopt a diagnostic approach that focuses on identifying problematics that hinder collective action and understanding how the problematics are constituted by different contextual factors (Ostrom, 2007; Ostrom et al., 2007). A diagnostic approach does not guarantee solutions; it, however, highlights the importance and perhaps the potential of trial and error and of being strategic in designing assistance schemes.

Third, the search for policy panaceas is largely based upon a static conception of institutions, focusing on efficiency as the measure of success. The whole search process has surrounded the question of what kind of institutional designs could lead to better irrigation performance. Relatively little attention has been given to the questions of whether and how institutions might generate or change the dynamics that underlie productive working relationships, how institutions might evolve in response to changes in the broader context, and what can be done to facilitate the emergence or development of particular institutions. As scholars who are deeply concerned with the potential of institutions for human development have long argued, working institutions are not mere words on paper but linguistic entities underlined by ideas and visions, and acted upon by a community of individuals who share common understanding of the meanings of the entities. Institutions are ongoing concerns; one could hope to introduce institutional change only if the complex dynamics of institutional change are taken seriously.

Fourth, while it is almost a truism to suggest that knowledge matters in the process of intervention, how to put in place effective knowledge generation and utilization mechanisms has often been assumed away in many intervention projects. It is true that the design of any intervention project needs to be built upon knowledge pertaining to both technological and social aspects of the intervention. As an intervention develops and evolves, information about its operation and how it affects the operation and dynamics of the SES affected, slowly becomes available. Whether actors in the system are able to gather, process, and utilize the information for continuously adjusting to the changing environment is fundamental to the sustainability of the positive impacts of an intervention effort. Successful interventions are not about bringing in benefits to an SES from the outside, but enhancing and sustaining the ability of the actors in the SES to learn and to utilize knowledge for problem solving.

SUSTAINABLE INTERVENTION: AN ANALYSIS OF AN INNOVATIVE INTERVENTION IN NEPAL

As a country that has received substantial amounts of aid from international donors over the last several decades, Nepal is among the countries that have the most intervention experiences. Particularly in irrigation, international donors and the Nepali government have invested large sums in the past decades in an effort to improve the country's irrigation performance. The external investments in irrigation systems in Nepal, however, have not fared well. The National Planning Commission of Nepal prepared a frank appraisal of the performance of irrigation development in Nepal from the mid-1950s through the mid-1990s when external assistance for building or rebuilding irrigation systems was substantial. The report stated that irrigation 'development and operation in Nepal is performing dismally relative to the amount of resources poured into this sector' (HMG/N, National Planning Commission of Nepal, 1994, p. i).

Many of the irrigation intervention projects in Nepal have been designed and implemented from an engineering-centered approach. As we have elaborated in earlier sections, such an approach focuses on putting in place organizational deliverables – engineering works, particular amounts of money, or particular institutional blueprints – as the means to improve irrigation performance.

To give the reader an idea about how such an approach operates, we present in Table 1.1 the implementation protocol of the Second Irrigation Sector Project (SISP), one of the largest irrigation intervention projects that have been implemented in Nepal. The protocol is mainly comprised of 13 steps (or components). These steps represent a sequence of imperatives that government officials were required to put in place in order to attain successful implementation. Perhaps because the list was made by the officials who were given the mandate to implement the intervention project, the 13 steps look like a checklist for making sure every 'deliverable' that was considered to be essential was in place. One interesting feature of the protocol is that while many of the steps seemed to involve farmers in the intervention process, the exact roles that the farmers

Table 1.1   The 13 steps for the implementation of the SISP

Step	
1	Water resources assessment (experiences of farmers are useful)
2	Information dissemination (interaction with farmer groups)
3	Farmers' request/application (request to be made by the farmers)
4	Identification (subproject identification)
5	Feasibility study (field-based information to be provided by the farmers)
6	Appraisal/approval
7	Establishing formal WUA (farmers' active role)
8	Survey and design; preparation of detailed cost estimates and tender documents (farmers need to know how much they are to contribute)
9	Memorandum of agreement (farmers' contribution of 15% of cost)
10	Construction (supervision for quality control by WUA members)
11	Commission (certification of performance by WUA)
12	O&M (regular operation and maintenance by the farmers)
13	Monitoring and evaluation

Source: Compiled from HMG/N, Department of Irrigation and ADB/Manila (1997, pp. 8–16).

played were rather limited and passive. Farmers were informed of what the government was planning to do, were asked about what they wanted, were told and trained to organize a WUA, and were told to contribute. They were, by and large, seen as beneficiaries of the intervention project whose major role was to provide input to fine-tune and facilitate the government's implementation of the intervention project.

In 1985, the Water and Energy Commission Secretariat (WECS) of Nepal and the International Irrigation Management Institute (IIMI) did develop an ingenious intervention program for 19 irrigation systems located in the central hills of Nepal. The project tried to overcome the 'best-practices' trap that prevailed at the time in regard to assisting irrigation systems (Yoder, 1986, 1994; P. Pradhan, 1989a, 1989b). The WECS/IIMI project was innovative in at least seven ways:

1.   The farmers could choose whether to be involved or not.
2.   The project provided technical assistance, but purposively did

not provide full funding for engineering improvements and the farmers were expected to provide core labor and some materials.

3. The farmers had to provide a full rank ordering of the improvements that they desired.

4. The farmers examined the engineering plans and had to OK them before they were implemented (in other words, the farmers had a veto over engineering plans that were not consistent with their preferences).

5. If the farmers were able to reduce the monetary expenditures for the highest-ranked projects by their own contributions, the released funds were then allocated to the next ranked project on the farmers' lists.

6. Participating farmers were expected to go through 'farmer-to-farmer' training offered by some of the more productive irrigation systems in Nepal.

7. Each farmer group was expected to write its own internal set of working rules that covered how future decisions would be made for their system.

In this book, we report intensively on this innovative intervention and examine how it was designed in a sincere effort to attain the sustainability of the effect of the intervention. Our aim is not simply to add another case study of intervention to the extant literature. Instead, we take the study of this innovative intervention as a window to address a fundamental question: how can one design interventions to help people in poverty without falling into the 'best-practices' trap and/or the 'dependency' trap? Drawing upon data collected across a time span of more than 15 years, this study will examine how the design of interventions could make a big difference not only in performance but also in the long-term sustainability of the impact of an intervention. We will identify key factors that help explain the performance of interventions, and explicate lessons that this intervention experience has for resource management and the management of development assistance.

The book is organized as follows. Chapter 2 will discuss broader issues related to interventions in Nepal and elsewhere so as to reflect on the applicability of the intervention initiative taken at a small watershed in Nepal for wider policy implications on donor assistance for community resource management. Chapter 3 will provide the background information of the intervention that this study focuses on. In particular, the chapter will focus on instances of innovative suggestions coming from farmers and how they were integrated in the implementation strategy. Chapter 4 will evaluate the long-term performance and sustainability of the intervention project using both longitudinal case studies and Qualitative Comparative Analysis (QCA). In particular, the analysis will focus on identifying factors that affect sustainability, and deciphering the ways these factors configure to affect performance. Chapter 5 will focus on post-intervention dynamics with detailed analysis of variables identified in previous chapters from success and failure cases. Also, the discussion will further focus on post-conflict opportunities and constraints. Chapter 6 will conclude by summarizing lessons we have learned from this study and their implications for designing irrigation intervention.

2 Effects of different modes of assistance on the performance of farmer-managed irrigation systems in Nepal

It is well documented that thousands of irrigation systems in Nepal are managed by the farmers themselves and that some of these farmer-managed irrigation systems (FMIS) have been in operation for centuries. Moreover, it has long been accepted by policymakers and donors that financial assistance is crucial in helping the farmers to construct permanent diversion structures, to line key parts of a canal and to undertake other capital-intensive work in order to improve the technical efficiency of these irrigation systems.

Furthermore, examples of recent interventions exist where assistance programs have been designed to improve institutional capacity of these local irrigators through enhanced social capital as well as effective self-governance mechanisms. Consequently, a number of different aid programs or interventions that aim to enhance FMIS performance by improving their physical infrastructures have been undertaken in Nepal. Despite the similar objectives of the intervening agencies, however, the consequences of these interventions have varied substantially. Given the increasing emphasis on the importance of intervention in improving irrigation performance, it is of great concern to assess why there is a variation of performance associated with the different types of intervention.

In this chapter, we will first briefly overview the history of irrigation development in Nepal and will describe the agencies that are involved in the extensive interventions as well as the processes of intervention. In the next section, we will discuss the rationale of a study of 229 irrigation systems in Nepal and some methodological procedures used in this study. Next, we will discuss the findings of this large-N study focusing on the factors that affect the performance of the irrigation systems in relation to the various interventions. In the final section, we will address the issues that require action from the intervening agencies in order to enhance irrigation performance.

In subsequent chapters, we then examine one specific type of intervention by WECS/IIMI and assess whether the said issues, which we identify in this chapter as requiring action from intervening agencies, have been addressed. If they have been addressed, how have the complexities been dealt with? Has due consideration been given in light of the various needs of the different irrigation systems' interventions? And finally, have these irrigation organizations sustained themselves over time in the dynamic world?

IRRIGATION DEVELOPMENT IN NEPAL

Irrigation development in Nepal appears to be as old as the history of agricultural development in Nepal. Irrigation systems in early times were constructed by local princes, or their officials, as well as by the farmers themselves. Much of the early evidence about irrigation development in the country is found in relation to taxation, land tenure, and customary laws. A feudal system of land ownership prevailed over much of Nepal until 1951. During this era, the country was ruled by the Rana dynasty with fundamental interests in collecting revenue and maintaining law and order.[1] When the Rana regime was overthrown following a people's revolution, Nepal began to be transformed from a medieval oriental despotism into a modern nation state.

Planned modes of irrigation development were initiated in Nepal with the establishment of the Department of Irrigation (DOI) in 1952. However, it was only with the implementation of the First Five-Year Plan in 1956 that significant efforts toward irrigation development were made by the government. Since then, requisite institutional arrangements from the center to district levels have been provided to the farmers who require the use of natural water resources for irrigation. A series of development plans and policy reforms have been implemented, including Basic Needs (1988), Water Resources Act (1992), Irrigation Policy (1992), and the Ninth Plan (1997–2002).

The Irrigation Policy of 1992 (revised in 1997 and 2002) has been instrumental in institutionalizing users' participation in the operation and management of irrigation schemes.[2] Two significant action plans spell out the concept of the Participatory Management Program of the DOI that emerged from this policy: (1) turnover of DOI-controlled irrigation schemes to the organization of beneficiaries and (2) joint management of large-scale irrigation systems. Implicit in both of these plans is the decentralization of responsibilities that would attract users' participation in the operation and management of irrigation systems, thus reducing the financial burden on the DOI. Although it has gone through several changes over the years in terms of its nomenclature, the DOI is the principal government entity involved in planning, developing, and managing government-owned irrigation schemes in Nepal.

Interventions in the Nepalese Context

Due to inadequate irrigation facilities, Nepal's agriculture continues to depend largely on rainfall, and FMIS continue to contribute significantly to the development of agricultural systems in Nepal. More than 70 percent of the total irrigated area of the country is served by FMIS alone (Gautam et al., 1992). The size of FMIS varies from less than 1 ha to as large as 15000 ha providing the irrigation needs of individual farmers (Yoder and Upadhyaya, 1987). FMIS, which have performed typically on a self-help basis in the past, now have an organizational basis for carrying out irrigation operation and management activities such as acquisition, allocation, distribution, resource mobilization, and conflict resolution. Regarding performance, FMIS have performed relatively better than most of the agency-managed irrigation systems (AMIS) (Lam, et al., 1994; Lam, 1998).

Given the agrarian economy of Nepal, where more than 90 percent of the population depends upon agriculture (Gautam et al., 1992), the government has regarded improving irrigation management as of major importance. Despite the high priority given to irrigation in each development plan, trends show that irrigated land area in Nepal has continued to decrease since 1993 (NPC, 1997). Moreover, while the potential of FMIS is substantial, not all FMIS are operating at an optimal level. Obviously, there is still much room for improvement in many systems. As a result, the government and some international donor agencies have implemented various programs to assist FMIS in Nepal over the last several decades.

There exist three categories of irrigation management in Nepal. The first category is governed and managed by non-users (by a government agency), the second is controlled and managed by water users (farmers) themselves, and the third is classified as joint management that aimed at initiating and promoting shared operation and management of large-scale irrigation systems between the DOI and water users. In this chapter, we will focus primarily on the second category of management and on diverse programs to assist FMIS.

After years of neglecting FMIS, recent policy developments in Nepal have focused on policies that have the avowed purpose of improving FMIS through various forms of intervention. Six broad sets of interventions that have had more than a few instances in recent history in Nepal are identified here: Department of Irrigation (DOI) through its field offices, Ministry of Local Development (MLD), multiservice agencies, Irrigation Sector Programs (Irrigation Line of Credit [ILC, World Bank] and Irrigation Sector Support Project [ISSP, Asian Development Bank (ADB)]), Agricultural Development Bank of Nepal (ADB/N), and Water and Energy Commission Secretariat (WECS). A description of these agencies and their processes of intervention is presented below.

Irrigation Development Division/Irrigation Development Subdivision (IDD/IDSD)
The District Irrigation Office (DIO) under the Department of Irrigation was mainly responsible for providing assistance to FMIS until 1999. After 1999, that responsibility fell to the Irrigation Development Divisions, which were established in the place of District Irrigation Offices and were in charge of two or more districts. The DIO used to intervene in the operation of FMIS by rehabilitating, extending, improving and constructing a new irrigation system either through its own implementing staff or by the farmers under their supervision. The extent of assistance is on the basis of financial resources available. Under this mode of intervention, the farmers, when their system has been identified, are not necessarily consulted in the feasibility study carried out by the DOI technicians or the consultants they hire. Construction work is generally undertaken by the contractors, and lukewarm attempts are made to form a users' committee.

Ministry of Local Development (MLD)

The Ministry of Local Development (formerly Ministry of Panchayat and Local Development) allocated funds for village-level projects through the then District Panchayats (now District Development Committees). The Ministry of Local Development established a Department of Local Infrastructure Development and Agricultural Roads (DOLIDAR) in 1998, which is to be the technical arm to all District Development Committees (DDCs). Small irrigation projects' improvements on the recommendation of District Agriculture Committees are being financed through DDCs. The farmers could request a DDC for the approval of their projects. Once a project was approved, the district administrative office would ask the farmers to form a construction committee and a formal contract would be signed between the committee and the district administration. After the release of the initial 50 percent of the estimated cost, construction work would start under the supervision of the district technical office. Additional funds would be released based on the progress of the work as certified by the technicians. After the project was completed, the construction committee was dissolved and a users' committee usually formed.

The availability of resources from the MLD was mainly based on the political influence of the area and local politicians, not on the basis of the needs assessment and improvement requirement. Some schemes received regular annual support while others did not receive any support at all. Neither the MLD and DDC, through which the funds were to channel, was equipped to undertake innovative measures in the intervention for FMIS improvement.

Multiservice agencies

Besides the above agencies, others were also involved in irrigation development in the country. These included the Farm Irrigation and Water Utilization Division (FIWUD) under the Department of Agriculture and several non-governmental organizations such as the International Labour Organization (ILO), United States Agency for International Development (USAID), Hill Food Production Project (HFPP), and United Mission to Nepal (UMN) as in the Andhi Khola and Agricultural Development Bank and CARE (Cooperative for American Relief Everywhere)/Nepal irrigation programs. These agencies undertook projects requested by those influenced by political lobbying as well as the farmers themselves through village assemblies. Usually, the construction work was carried out by forming a construction committee, which was converted into a users' committee after the project was completed. The completed project was then handed over to the users' committee for operation and management. Interventions by most of these agencies aimed at overall agricultural development inclusive of irrigation development.

In particular, the FIWUD intervention included two components: (1) helping better utilization of irrigation water at tertiary and field channels of large-scale irrigation systems and (2) helping the farmers in constructing, improving, and maintaining their own irrigation systems by making optimum use of available water for increasing agriculture production. The FIWUD intervention was initiated in 1973 and by 1986 it provided assistance to 106 systems in hill and *terai* covering 19 600 ha. The FIWUD intervention included the following important elements:

- application by the beneficiaries for assistance with the endorsement of District Agriculture-Irrigation Committee of Panchayat District;
- rapid appraisal of the candidate system by FIWUD officials;
- provision of technical assistance related to designing and evaluating the system as well as managing construction and supervision;
- use of simple technology;
- beneficiaries' involvement at all stages of planning and implementation;
- promoting self-help attitude among the farmers;
- FIWUD's program of assistance to irrigation systems extends all over the country;
- cost-sharing mechanism in which the government provides grants to cover 75 percent of the cost of materials and construction of the structures, while the beneficiaries contribute the remaining 25 percent; and
- cash contribution amounting to 5 percent before the implementation of the program.

In 1989, FIWUD was amalgamated with the DOI, but the DOI did not incorporate the stated FIWUD assistance features into its assistance program.

Irrigation Sector Support Project (ISSP)

In the late 1980s, the approach for investment changed from project-specific to irrigation sector-specific. For irrigation sector implementation, the World Bank took responsibility for 40 districts in the western, mid-western, and far western regions and ADB took responsibility for 35 districts in the central and eastern regions of Nepal. There are two major investment programs under the ISSP: the Irrigation Line of Credit (ILC) and the Irrigation Sector Project (ISP). The Department of Irrigation operates mainly through these programs.

Launched in 1989, the ILC program received technical assistance from the United Nations Development Programme (UNDP) and loan assistance from the World Bank in 34 districts of the Western Development Region that utilizes water exclusively from rivers and streams. On the other hand, the ISP was started in 1989 to facilitate the government's efforts in terms of investment in irrigation development in a sectoral approach. The program was initiated with loan assistance from the ADB, and aimed to provide irrigation facilities in 35 districts of the Eastern and Central Development Regions of Nepal. The ISSP mainly carried out three types of programs: (1) support of FMIS in the private sector, (2) turnover of government-managed irrigation schemes to an association of beneficiaries in the private sector and (3) participatory or joint management of large irrigation schemes.

In the private sector, ISSP provided irrigation assistance in a 'demand-driven development' concept for field implementation. This meant that the project targeted only those programs that were initiated and requested by water users. For this, a formal request from an organized group duly signed by more than 66 percent of the potential beneficiaries formed the main basis upon which the agency initiated activities to implement a program. This concept was followed both in the ILC and ISP programs. The programs were approved after carrying out a preliminary feasibility survey by a team of technicians from the DIO and the beneficiaries. The programs that were approved were then implemented with the involvement of the DIO. Such programs were implemented with the concept of participation of the organized water users. The water users were involved in all stages of irrigation development, from program selection to design and construction activities, including disbursement of funds for field operations.

In terms of cost sharing, water users were required to bear 1 to 5 percent cash plus contribute labor equivalent to 6 to 20 percent of the total cost. The balance of the cost was met by the ISSP. One of the basic tenets of this program was that the implementing agency would ensure that at least 20 percent of the executive members of the organized beneficiaries were females from the community. Upon the completion of the programs, the irrigation systems were formally handed over to the organization of water users for operation and maintenance. It was the effectiveness of operation and maintenance by organized water users that determined the long-term sustainability of irrigation systems rather than the amount of the initial investment of the government.

The program's activities included rehabilitation and improvement of FMIS in the command area and construction of new small- and medium-scale gravity-flow irrigation schemes that served the irrigation needs of a significantly large proportion of the irrigated areas in the country. The program also included strengthening the institutional capacities of DIOs and Regional Irrigation Directorates to support and institutionalize the participatory irrigation development and management programs under ISSP.

One of the projects in this category was the SINKALAMA Project, funded by ADB/Manila for improving FMIS in hill districts of Sindhupalchok, Kavrepalanchok, Lalitpur and Makwanpur in 1988. The SINKALAMA Project (so-called by combining the initial letters of the project districts), with the participation of the farmer-users, sought to improve the physical structures of the FMIS. The implementation responsibility was taken by the DOI through its District Offices.

This assistance became a one-shot activity. As long as the fund for improvements was available, the farmers were active. But when the fund dried up, the farmers became inactive and, eventually, the DOI that was responsible for administering the project implementation also became inactive. Fifty-five irrigation systems commanding 2500 ha received support for improvements. Institutional strengthening support, however, was weak and thus the long-term sustainability of the system was at risk. In addition, the project provided no support to strengthen the Water User Association (WUA). For example, the post-project evaluation indicated that 90 percent of the construction committees, which were committees that were supposed to be converted into regular management committees after the completion of the construction, did not convert into management committees (Shrestha, 1991; Karki, 2001).

Agricultural Development Bank of Nepal (ADB/N)

Various donor agencies have also implemented intervention programs. Among these programs, the Agricultural Development Bank of Nepal (ADB/N) has demonstrated the possibility of supporting institutional development through credit and subsidy programs for the development of small-scale irrigation facilities. The ADB/N has also collaborated with CARE/Nepal to provide a subsidy in the form of material support to several irrigation systems. The ADB/N has supported the development of about 106 000 ha where these systems are mostly user-controlled (P. Pradhan, 1989a, p. 2). The ADB/N's program includes surface irrigation, shallow and artisan tube wells, installation of a low-cost manual lift, and mechanical irrigation pumps.

ADB/N started the Community Surface Irrigation Project (CSIP) as a regular program on the basis of community participation since 1983/84. Before the selection of the irrigation project, farmers would be informed by the Group Organizer of ADB/N (often working as a Group Organizer for the Small Farmer Development Program within the area of the irrigation system) about the rules of securing support from the ADB/CARE program for the improvement of the irrigation system. In the case of the ADB/CARE Nepal program, farmer users have to contribute 40 percent of the cost of the system improvement. This contribution takes the form of a group loan and labor contribution. The group organizer becomes the key person to mobilize the farmers for irrigation development and management. Presently, as in the case of government-sponsored irrigation programs, the ADB/N's irrigation program also operates fully on a demand-driven mode. In addition to the formal request, the beneficiaries are also required to submit land title certificates as collateral against the loan to be advanced to them.

As a process, ADB/N's intervention starts with project identification through a Small Farmer Development Project Baseline Survey. The perceived needs of the farmers are then prioritized. When irrigation gets top priority, ADB/N conducts a feasibility study and, if feasible, an irrigation group is formed or identified if it already exists. The group is expected to contribute 10 percent of the cost as labor, 60 percent is funded as a government grant, and ADB/N provides

30 percent as a loan to the group. Technical assistance is provided by ADB/N and construction is usually carried out by the farmers, or if there is no local expertise, small-scale contractors are used. On completion, ADB/N hands the project over to the irrigation group, but continues to provide technical services if needed.

Water and Energy Commission Secretariat and the International Irrigation Management Institute (WECS/IIMI)

The Water and Energy Commission Secretariat (WECS) of Nepal, with assistance from the Ford Foundation and the International Irrigation Management Institute (IIMI)[3] initiated an action-research project in 1985 in the Indrawati watershed basin in the Sindhupalchok District. The project objectives were to establish low-cost procedures for identifying the needs of farmers in a larger area and to develop and test methods for delivering assistance that improved the capacity of FMIS in Nepal (WECS/IIMI, 1990, p. 12). Researchers associated with IIMI identified 23 irrigation systems in the watershed where expansion of the system was potentially feasible and existing users of the system were initially willing to allow expansion and to accept additional farmers as members of a water users' organization (Lam and Shivakoti, 2002).

The farmers in these systems were involved in all aspects of the planning and operation of the project. After extensive discussions, the farmers from 19 systems agreed to participate. The farmers knew from the beginning that the budget to be used to help support expansion would be modest and that they themselves would need to do most of the construction. The farmers were asked to list improvements that they desired and to make a rank-order priority of them. The improvements that were most necessary for the expansion of the system, but difficult for farmers to do without financial assistance, were to be given highest priority. The farmers were then assigned a firm budget for the project and were told that if they could save money on the first-priority work, they could use it for the second- or even the third-priority work. The intention was to create a positive incentive for the farmers to use project funds with the greatest of care.

Another important aspect of this project was 'farmer-to-farmer' training. Groups of farmers from the 19 systems were taken to farmer-managed systems that were known to be very efficient and well organized (Yoder, 1991a). The host farmers from the effective systems described the ways they had dealt with core organization

issues such as labor mobilization for routine or emergency maintenance, water allocation and distribution, conflict management, and
the structure of their organizations (N. Pradhan, 1987). The trainee
farmers also were shown the constitution and the minutes and
attendance records taken at meetings and at sessions where labor
was mobilized to maintain or repair the systems. Thus, in addition to
this intervention being demand driven, the WECS/IIMI intervention
paid serious attention to developing incentives that would involve
the farmers in making good long-term investments in their own
systems, and providing them with the kind of training they needed
to improve their own organizational structure and operating procedures. Since the process of intervention is innovative, has several
features that have unique characteristics, and has been successful in
sustaining the performances of systems assisted, we have analyzed
these systems in detail in Chapters 3–5 of this book.

THE RATIONALE AND METHODOLOGY OF
INTERVENTIONS IN NEPAL

As described in the previous sections, a wide variety of differently
designed interventions have been implemented by diverse programs
within government and other agencies frequently supported by
donor agencies. All of these programs have the intention of increasing agricultural productivity by developing and improving existing
FMIS. However, the efforts of the government of Nepal to assist
FMIS have not been uniformly successful. While some interventions
have enabled farmers to maintain their irrigation systems at a lower
cost and increase their overall efficiency, others have had a damag-
ing effect. In some systems, the organizational effectiveness of some
irrigation institutions has declined; farmers who used to maintain
their systems regularly no longer do so; disputes over water rights
have increased; and, in some instances, the total land irrigated and
the yields obtained have decreased after interventions that were
intended to increase them (Shivakoti et al., 1992; Shivakoti and
Ostrom, 2002). In essence, these interventions vary substantially
in regard to the extent of farmer participation in their design and
implementation.

The above variance in results is a puzzle that needs to be resolved.
It is important to learn from past successes as well as failures so

as to improve future policies. It is thus important to analyze and understand how various modes of intervention intended to improve the performance of FMIS resulted in high variance in levels of performance. What could be the physical and institutional factors that affect the performance of irrigation systems? Are there systematic differences in how diverse intervention strategies are initiated and planned? In general, what lessons can be learned to increase the probability of successful future interventions? These are some of the questions we will now seek to address.

We speculate that there is a difference in the success of interventions that are initiated by the water users themselves as contrasted with those interventions that are carried out without fundamental consultation with the farmers. Among the types of interventions discussed above, the program that involved the farmers to the greatest extent was the one that was undertaken by WECS/IIMI. Most of the interventions organized by the Agricultural Development Bank of Nepal with CARE funding and the ISSP program also relied heavily on farmer involvement in the initial planning and implementation of the intervention.

On the other hand, many of the DOI, MLD, and multiservice programs are strongly 'supply driven.' Thus, we categorize these six types of interventions into three broad groupings: (1) interventions that used a demand-driven approach with a high level of farmers' participation, (2) interventions that were primarily demand driven but with a moderate level of farmers' involvement, and (3) interventions that were primarily supply driven by the agency undertaking and/or with a low level of farmers' involvement in the effort to improve irrigation systems. In the first group, we would place the WECS/IIMI project. In the second group, we would place ADB/N and ISSP. In the third group, we would place DOI, MLD, and multiservice agencies. We would expect to see that the more demand-driven programs are somewhat more successful than the supply-driven intervention programs.

We also seek to examine how various physical, social, and institutional variables affected the performance of different types of FMIS depending upon the type of intervention they have received as well as FMIS that have never received an intervention. To do this, we have used the information in the Nepal Irrigation Institutions and Systems (NIIS) database that has been developed by colleagues associated with the Irrigation Management Systems Study Group at the Institute of Agriculture and Animal Science (IAAS), Tribhuvan University, Nepal, in collaboration with colleagues at the Workshop in Political Theory and Policy Analysis, Indiana University, USA.

Specifically, the NIIS database has been developed over a substantial period of time. The first systems coded in this database included all of the case studies written about individual irrigation systems in Nepal that could be found by scholars at the Workshop in Political Theory and Policy Analysis. We located 130 case studies that had substantial information about the physical structures and social organization of irrigation systems and coded as many variables as possible for 127 of these from written records. Three cases turned out to have insufficient data for final coding (see Ostrom and Benjamin, 1991).

Due to the problems associated with missing variables, we asked the Ford Foundation to support an effort to visit many of these systems to complete the missing information and to check on the information contained in the original coding. At the end of this process, we had identified 150 systems. An initial analysis was conducted by Wai Fung Lam (1998), which showed that FMIS in Nepal performed at a substantially higher level than AMIS. We were urged to add to the original NIIS database so as to address several competing hypotheses related to the age and size of the systems. With the additional systems added, the original findings were sustained (see Appendix C in Lam, 1994).

During the years since 1993–97, colleagues at IAAS in Rampur have added 86 additional systems. One group of systems was located in the Chitwan District and an effort was made to collect information about these systems before they received help and assistance from the East Rapti Irrigation Project. Consequently, we now have information about 231 systems spread out across 29 districts of the 75 districts in Nepal. For 24 of these systems, we have information collected at two different time periods.[4] In this analysis, we have utilized the data collected in the second time period.

The irrigation systems coded in the NIIS database are not a 'random sample' of the irrigation systems in Nepal. No one knows how many irrigation systems there are and no one has any kind of list that could be used as an initial sampling frame. To our knowledge, however, this is the largest collection of information about irrigation systems in one country that exists anywhere in the world. We initiated this project with the hope that we could learn a great deal from any effort to do a quantitative, rather than a strictly qualitative, analysis of the factors affecting irrigation system performance.

EXAMINING THE PERFORMANCE OF DIVERSE TYPES OF INTERVENTIONS

As discussed above, three different types of strategies are adopted by agencies involved in the intervention of the 114 FMIS included in this study. The period of intervention described here dates back to as early as 1957 by the Department of Irrigation, with the most recent ones in 1997. About 21, 40, 28, and 11 percent of these irrigation systems are located in the hill-river valleys, hills, *terai*-river valleys, and *terai*, respectively. The area of farmland served by these systems varies from a minimum of 4 ha to a maximum of 9816 ha, with an average area of 263.4 ha. The total canal length of these systems varies from 0.425 to 79.5 km, with 7.1 km as the average length. Most systems (about 76 percent) lack access to alternative sources of water supply.

In terms of age, there are systems built as early as in the sixteenth century with recent ones built in the late 1990s. The number of water users varies from five in a particular system to 2500 (the average is 207 users). In general, water users are quite heterogeneous in regard to socioeconomic status. Except for WECS/IIMI, which had projects only in the hills, all other agencies have intervened in irrigation systems located in the hills as well as *terai*. Each irrigation system has one WUA that is entrusted with allocation and distribution of water, conflict resolution, and maintenance of canals and headworks. Obviously, given the heterogeneous sociophysical attributes of the irrigation systems included in this analysis, the management tasks of the agencies involved in the intervention program will have differed considerably.

Since we do not have before–after data for most of the 114 FMIS that received some external assistance, we compare these systems with the performance achieved by two other types of systems: (1) FMIS, which have not received external assistance and (2) AMIS, which have been constructed and operated by the Department of Irrigation. These two comparison groups give us different information. The FMIS provide a comparison with a group of irrigation systems that are entirely demand driven. All improvements to these systems have been designed and undertaken by the farmers themselves. The AMIS are entirely supply driven since the DOI determines where they will be constructed, how they will be operated and the level of maintenance supplied.

The comparison among the intervening agencies has been made in relative terms with respect to the performance of irrigation systems that have undergone one or another mode of intervention. Specifically, the discussion of performance of irrigation systems includes three dimensions. These are: the physical condition of irrigation systems, water delivery, and agricultural productivity as three coherent parts of irrigation performance. These three dimensions are not additive nor can any one of them be completely substituted by any of the others. Rather, they are interdependent on one another. An irrigation system cannot be said to perform well if its canals are well maintained but its water delivery is unsatisfactory. Similarly, the performance of an irrigation system is problematic if, with effective water delivery, farmers are not able or encouraged to use the water efficiently to increase agricultural productivity.

Physical Condition of Irrigation Systems

The physical condition of irrigation systems pertains to whether the system provides a sound technical basis for effective water delivery. This dimension consists of two aspects. The first is the technical efficacy of irrigation infrastructures. Technical efficacy of a system is the capacity of a system to deliver water from headworks to outlets. It is concerned with whether or not the infrastructure is well maintained. A technically efficient system is one that minimizes water loss in the process of delivery (Sparling, 1990). The technical efficacy of a system is determined by its physical characteristics, such as types of headworks, terrain, and canal lining.

The second aspect is the economic efficacy of maintaining irrigation systems. The focus in this instance is on the cost–benefit calculus of maintenance. In many rural areas of Nepal, resources are often scarce. The time, effort, and money spent on system maintenance could mean a significant investment. The more resources farmers use in maintenance, the fewer are left over for agricultural activities. This, then, has a direct bearing on farmers' incomes.

The result from the analysis of the condition of irrigation systems under various interventions is presented in Table 2.1. This table

Table 2.1   Relationship between intervention type and physical condition of irrigation systems

Types of intervention	Physical condition			
	Excellent	Moderately good	Poor	Total
Intervention with high levels of farmers' involvement (WECS/IIMI)	12 (63.2)	7 (36.8)	0 (00.0)	19 (100.0)
Intervention with moderate levels of farmers' involvement (ADB/N and ISSP)	9 (13.6)	45 (68.2)	12 (18.2)	66 (100.0)
Intervention with low levels of farmers' involvement (DIO, MLD, and multiservice agencies)	6 (22.2)	16 (59.3)	5 (18.5)	27 (100.0)
FMIS without intervention	6 (8.7)	54 (78.3)	9 (13.0)	69 (100.0)
AMIS	4 (8.4)	22 (45.8)	22 (45.8)	48 (100.0)
Total	37 (16.2)	144 (62.9)	48 (20.9)	229 (100.0)

Note: The figures in parentheses indicate percentage.

Source: Authors.

shows a substantial degree of variation in the overall physical condition of irrigation systems. The physical structures of the WECS/IIMI-assisted systems outperformed all the systems supported by other agencies. Nearly two-thirds of the systems were rated as in excellent physical condition for the basic construction design of the system. No other interventions came close to this. It must be noted that all the WECS/IIMI-assisted systems had temporary (brushwood or sandstone) types of headworks and were partially lined. In contrast, a majority of FMIS that had received external support with a moderate level of farmers' participation or with a low level of input were in moderately good or poor condition. Many of these systems did have a permanent type of headworks (gabion boxes, concrete weirs, barrages, etc.).

It is noteworthy that AMIS, about 83 percent of which had permanent headworks, had less than one-tenth of the systems that were excellent in terms of physical condition. It is also the case, as will be shown later (Table 2.5), that AMIS were less able than other types of irrigation systems to get adequate and predictable water to the tail end of their systems even though most of them have permanent headworks and are fully lined. Lam (1998) reasons that the existence of permanent headworks often exacerbates the asymmetries between headenders and tailenders, thus leading to low levels of performance of systems. In such an asymmetrical situation, the tailenders are not likely to take good care of the physical structures of the systems, nor follow the rules concerning water distribution and allocation.

In terms of the technical efficiency of systems, Table 2.2 shows that a majority (about 60 percent) of all of the systems were found to be only moderately efficient. This implies that, given the other constraints the farmers face, a considerable loss of water occurs between the head end and the tail end of the system in most systems. Similarly, in terms of economic efficiency, most systems are just moderately efficient. An economically efficient irrigation system is one where the cost of operating and maintaining the system is less than the benefits obtained from operation and maintenance.

When looked at across various categories of intervening agencies, WECS/IIMI-supported systems appeared to be technically and economically much more efficient compared with others (Tables 2.2 and 2.3). On the other hand, a higher proportion of the systems with relatively low levels of farmers' involvement in the irrigation projects appeared to be technically and economically less efficient.

The WECS/IIMI strategy of farmer-to-farmer training programs during the annual general meeting of the WUA may have contributed to a higher level of physical performance shown by the irrigation systems. The training program aimed to improve the capabilities of water users for operation and maintenance of systems, after handing over of the project to the users. In most of the other intervention programs, the training component was lacking or those who did provide training did not organize this program in an effective manner. Keeping a system in good condition requires intensive labor mobilization. In Nepal, manual labor is extensively utilized in the operation and maintenance of irrigation systems. But in about

Table 2.2   Relationship between intervention type and technical efficiency of irrigation systems

Types of intervention	Technical efficiency			
	Highly efficient	Moderately efficient	Inefficient	Total
Intervention with high levels of farmers' involvement (WECS/IIMI)	14 (73.7)	5 (26.3)	0 (00.0)	19 (100.0)
Intervention with moderate levels of farmers' involvement (ADB/N and ISSP)	23 (35.4)	38 (58.5)	4 (6.2)	65 (100.0)
Intervention with low levels of farmers' involvement (DIO, MLD, and multiservice agencies)	6 (22.2)	19 (70.4)	2 (7.4)	27 (100.0)
FMIS without intervention	9 (13.0)	51 (73.9)	9 (13.0)	69 (100.0)
AMIS	6 (12.5)	24 (50.0)	18 (37.5)	48 (100.0)
Total	58 (25.4)	137 (60.1)	33 (14.5)	228 (100.0)

Note: The figures in parentheses indicate percentage.

Source: Authors.

one-third of the total irrigation systems, the rules and regulations pertaining to labor mobilization were found to be poorly enforced. When the rules are not well enforced, the tendency for water users to avoid contribution of voluntary labor in the operation and maintenance of systems becomes more prevalent.

It is interesting to note that a greater proportion of AMIS operated without farmers' participation were technically and economically inefficient as compared with the externally assisted systems with some level of farmers' involvement as well as the systems exclusively managed by the farmers themselves (Tables 2.2 and 2.3). This situation can also be viewed as a reflection of the rule-ordered

Table 2.3   *Relationship between intervention type and economic efficiency of irrigation systems*

Types of intervention	Economic efficiency			
	Highly efficient	Moderately efficient	Inefficient	Total
Intervention with high levels of farmers' involvement (WECS/IIMI)	16 (84.2)	3 (15.8)	0 (00.0)	19 (100.0)
Intervention with moderate levels of farmers' involvement (ADB/N and ISSP)	21 (31.8)	41 (62.1)	4 (6.1)	66 (100.0)
Intervention with low levels of farmers' involvement (DIO, MLD, and multiservice agencies)	9 (33.3)	18 (66.7)	0 (00.0)	27 (100.0)
FMIS without intervention	14 (20.3)	53 (76.8)	2 (2.9)	69 (100.0)
AMIS	6 (12.5)	25 (52.1)	17 (35.4)	48 (100.0)
Total	66 (28.8)	140 (61.1)	23 (10.0)	229 (100.0)

Note: The figures in parentheses indicate percentage.

Source: Authors.

relationships with regard to governance structure and institutional arrangement. Lam (1998) argues that AMIS might have all kinds of rules to be employed. Yet, these rules are perceived by water users on some AMIS as commands imposed by irrigation officials, which are to be worked around instead of to be worked with. When this happens, a set of rules is not likely to result in productive working relationships among the water users.

On the other hand, the rules are made by the farmers collectively in FMIS. This can facilitate the development of shared norms that emphasize the importance and viability of self-reliance and cooperation in dealing with collective action. In his study, Lam (1998) found the involvement of farmers to be more likely in FMIS, as compared with their counterparts in AMIS. They were also more involved in entrepreneurial activities (such as crafting rules, discussion of issues of common concern, activities that facilitate the organization of various collective actions concerning irrigation governance and management) in an attempt to achieve coordinated strategies pertaining to operation and maintenance of irrigation systems.

Water Delivery

Water delivery is concerned with problems of water distribution and allocation. The dimension of water delivery captures not only the adequacy of water delivery but also elements such as equity and reliability. Water adequacy refers to whether an irrigation system is able to make enough water available to meet the irrigation needs of the farmers dependent on a system. Often, farmers have seasonal variations in terms of demand for water. Thus, adequacy should be measured in terms of the needs of farmers, keeping in view the seasonal variation in their demand for water.

Equity pertains to the allocation of available water in an equitable manner so that farmers who need water to cultivate their land are enabled to do so more effectively; in other words, the allocation of water in such a way that farmers, irrespective of whether they are headenders or tailenders, would receive their fair share of water. Reliability refers to the predictability and timeliness of water delivery to farmers. Predictability implies the knowledge about fluctuation of water flow in advance, and timeliness means that the schedule of water delivery is appropriate in terms of the needs of the farmers.

The data in Table 2.4 illustrate that less than half (about 44 percent) of the total systems seemed to meet the irrigation needs of tailenders, with sufficient and predictable amounts of water. When viewed by the type of intervention, a big chunk (79 percent) of the WECS/IIMI-assisted systems delivered sufficient amounts of water, even to the tailenders as per the allocation schedule known to them. On the other hand, the water distribution pattern did not seem to be adequate in almost 70 percent of the systems that had a low level of farmers' involvement. In 15 percent of the systems, water was both inadequate and unpredictable. Severe problems can be noticed among AMIS where more than 85 percent of the systems appeared to deliver insufficient amounts of water to tailenders (and 36 percent

*Table 2.4   Relationship between intervention type and water supply
at tail ends of irrigation systems*

Types of intervention	Water supply at tail ends				
	Adequate and predict- able	Adequate and unpredict- able	Inad- equate and predict- able	Inad- equate and unpredict- able	Total
Intervention with high levels of farmers' involvement (WECS/IIMI)	15 (78.9)	1 (5.3)	2 (10.5)	1 (5.3)	19 (100.0)
Intervention with moder- ate levels of farmers' involvement (ADB/N and ISSP)	37 (58.7)	4 (6.3)	17 (27.0)	5 (7.9)	63 (100.0)
Intervention with low levels of farmers' involvement (DIO, MLD, and multiser- vice agencies)	7 (26.9)	1 (3.9)	14 (53.8)	4 (15.4)	26 (100.0)
FMIS without intervention	35 (50.7)	0 (00.0)	26 (37.7)	8 (11.6)	69 (100.0)
AMIS	5 (10.6)	1 (2.1)	24 (51.1)	17 (36.2)	47 (100.0)
Total	99 (44.2)	7 (3.1)	83 (37.1)	35 (15.6)	224 (100.0)

Note: The figures in parentheses indicate percentage.

Source: Authors.

of the systems delivered both insufficient and unpredictable amounts to tailenders).

It might be argued that while the lofty objective of the intervening agencies to expand the service area has been achieved, effective water delivery problems related to institutional arrangements of irrigation governance and management still remained unsolved. In fact, the seasonal variation in the availability of water also affects the distribution pattern at the head and tail ends of a canal. In all the systems, somewhat more water was available at head ends year-round, as compared with tail ends. We have also observed many irrigation systems where the water users had conflicts among themselves over the issue of water allocation and distribution in the absence of clearly defined organizational rules and regulations related to systematic allocation and distribution of irrigation water. Tang (1992) has pointed out that water allocation is a major source of conflict in irrigation and this problem has to be solved by some form of institutional arrangement.

The information presented in Table 2.5 reveals a variation in distribution of irrigation water across the systems under all intervening agencies. On the whole, in about 24 percent of the systems, some farmers were either deprived of their fair share of water and/or given a lower priority in the distribution of water. This problem was more prominent in AMIS, where about 56 percent of the systems encountered such situations. In contrast, a minimum of the ADB/N and ISSP-assisted systems faced such problems. In this particular instance, only 10 percent of systems were in a disadvantageous position. The condition of inequality was more conspicuous in the supply-driven mode of intervention, as compared with demand-driven interventions. This implies an ineffective institutional arrangement on the part of the irrigation governance in regard to equitable distribution of water. It is highly likely that inequality leads to collective inaction owing to lack of incentives to cooperate on the part of downstream farmers, while head-end farmers enjoy more benefits.

Agricultural Productivity

The dimension of agricultural productivity connotes the productivity of the farmland served by a particular irrigation system. Agricultural productivity has been assessed in terms of cropping intensity. This places emphasis on the intimate relationships between the collective

Table 2.5   Relationship between intervention type and water distribution pattern

Types of intervention	Irrigation systems with disadvantaged farmers due to inequitable distribution of water		
	Yes	No	Total
Intervention with high levels of farmers' involvement (WECS/IIMI)	5 (26.3)	14 (73.7)	19 (100.0)
Intervention with moderate levels of farmers' involvement (ADB/N and ISSP)	6 (10.0)	54 (90.0)	60 (100.0)
Intervention with low levels of farmers' involvement (DIO, MLD, and multiservice agencies)	6 (22.2)	21 (77.8)	27 (100.0)
FMIS without intervention	9 (13.8)	56 (86.2)	65 (100.0)
AMIS	27 (56.3)	21 (43.8)	48 (100.0)
Total	53 (24.2)	166 (75.8)	219 (100.0)

Note: The figures in parentheses indicate percentage.

Source: Authors.

action of irrigation governance and management, and the local communities that the irrigation systems are supposed to serve. From the perspective of farmers, a high-performance irrigation system should be one that can increase their agricultural productivity and, hence, improve their livelihood. Although achieving lofty goals such as national economic development might be an important concern for the government, it is not likely to be the major concern of farmers when they contemplate undertaking collective action with their peers at the local level.

As revealed by Table 2.6, the pattern of productivity suggests a substantial variation in the cropping intensities across the systems

*Table 2.6   Relationship between intervention type and agricultural
productivity of irrigation systems*

Types of intervention	Agricultural productivity	
	Cropping intensity at head (%)	Cropping intensity at tail (%)
Intervention with high levels of farmers' involvement (WECS/IIMI)	252.42 (N = 19)	246.15 (N = 19)
Intervention with moderate levels of farmers' involvement (ADB/N and ISSP)	235.89 (N = 64)	229.10 (N = 65)
Intervention with low levels of farmers' involvement (DIO, MLD, and multi-service agencies)	247.40 (N = 25)	233.08 (N = 25)
FMIS without intervention	255.75 (N = 60)	251.03 (N = 57)
AMIS	211.54 (N = 46)	196.22 (N = 44)
Total	239.03 (N = 214)	230.18 (N = 210)

Note: The figures in parentheses indicate number of irrigation systems.

Source: Authors.

under different modes of intervention. The cropping intensity varied from a lowest of 100 percent (one crop a year) to a highest of 300 percent (three crops a year) at both head and tail ends of irrigation systems under various modes of intervention. However, the average cropping intensity at the head end was a little higher (238.28 percent) than that of the tail end (230.23 percent). When comparing the cropping intensities among various intervention modes, the WECS/IIMI-assisted systems excelled over all other systems. On the other hand, AMIS had the lowest performance. The difference in the cropping intensities at two different canal ends might be due to the fact that the upstream is generally more likely to have reliable and timely access to water as compared with the downstream. As mentioned earlier, many irrigation systems in this study also faced this sort of situation.

The comparison of externally assisted systems with AMIS and FMIS showed that a relatively higher performance in those systems had at least some degree of farmers' participation. The lower agricultural productivity of AMIS could be a reflection of the overall poor physical condition of irrigation systems and the pattern of water distribution. As we discussed earlier, a relatively higher proportion of AMIS had a poor overall physical infrastructure as well as facing an inequitable pattern of water distribution, thus hampering the users at the tail end.

Since many efforts to 'assist' irrigation systems attempt to increase the size of the command area, we wanted to examine the impact of the size of command area on cropping intensity. We performed a correlational analysis of 108 agency-assisted irrigation systems. The analysis resulted in a negative relationship between the area under irrigation and cropping intensity. More specifically, cropping intensity at head ($r = -0.215$, $p = 0.05$) was significantly but negatively related to the service area covered. This implies that while the agencies are concerned with achieving their target of increasing the command area, they should also pay attention to the equitable distribution of water among water users. This is because lack of symmetries in the water allocation and a distribution pattern may lead to deleterious effects on the agricultural productivity. Further, disadvantaged farmers often show a reluctance to be involved in the regular operation and maintenance of physical infrastructures of irrigation systems that require extensive manual labor.

CONCLUDING REFLECTIONS

In summary, the analysis of different modes, types, and levels of interventions have illustrated the disparity of performance of differently organized irrigation systems. This disparity in performance has a bearing on the effectiveness of interventions. While one particular mode of intervention has contributed substantially to the improved performance of irrigation systems, others have a somewhat poorer record. In general, the overall performance of interventions in irrigation systems in a demand-driven mode resulting from a higher level of farmers' involvement in irrigation projects has been better than those assisted in a supply-driven mode involving a moderate or low level of farmers' participation. In particular, the WECS/IIMI example provides an excellent model of intervention that appeared to have paid equal attention to both physical as well as social systems of WUAs.

These examples provide useful information about the potential of externally assisted irrigation schemes implemented in various modes and the factors that affect the performance of irrigation systems. Most intervention projects focus more on the physical infrastructure of irrigation systems, ignoring the social infrastructure. It is a fact that in the improvement of any irrigation system, the key role of a physical infrastructure cannot be denied. However, the social and institutional aspects have no less important role to play in the improvement of irrigation system performance. In the irrigation schemes implemented without consultation of the potential water users, it is very likely that the users tend not to contribute voluntary labor in the operation and maintenance tasks to be performed on a collective-action basis. The irrigation systems that require collective action on a regular and organizational basis tend to succeed when programs are implemented in a genuinely demand-driven style consistent with the concept of people's participation.

Given the subsistence nature of farming and financial constraints on most farms in Nepal, 'intervention' as a major strategy adopted by the government to improve the performance of FMIS is potentially of considerable help. However, as indicated by this study, emphasizing the improvement of physical infrastructures is not sufficient. Improving performance necessitates a more comprehensive approach, encompassing the development of both physical capital as well as social capital that provides complex systems of institutional arrangements. Mere improvement of physical systems cannot enhance the performance of irrigation systems.

In fact, the task of irrigation development goes far beyond the mere construction or rehabilitation of physical systems. The involvement of potential beneficiaries at all stages of irrigation development is inevitably important in order to manage social conflict and growing perceptions of social inequality in the allocation and distribution of irrigation water. In this regard, Uphoff et al. (1991) insightfully point out that focusing on irrigation management should not be considered only as a sociotechnical enterprise but also as an organizational-managerial one.

Viewed from this perspective, the organizational effectiveness of WUAs is indispensable to the management of irrigation systems.

Therefore, the improvement of physical infrastructures together with the development of social infrastructures should be an area of macro-reform that national governments, including intervening agencies, should take into consideration as a policy. While WUAs represent a major means of improving irrigation management by expanding farmer participation and responsibility, they are not always or everywhere effective without an improvement in the infrastructure of their systems (Uphoff, 1986).

NOTES

1. The Rana family came into political power in 1846 and continued the family prime ministership of Ranas until they were thrown out in 1951.
2. The Irrigation Policy is supposed to be revised every five years to make sure that the policy is adaptive to the changing environment. The process of revision for 2007 started, but because of political instability, the process could not be completed and is still pending.
3. In November 1997, IIMI changed its name to the International Water Management Institute (IWMI).
4. Of these, five systems are located in Dang District and were assisted by the ADB/N, and 19 systems are located in Sindhupalchok District and were part of the WECS/IIMI intervention described above. The written case descriptions for these systems were so detailed and accurate that when we visited these systems, we coded them as of the second visit rather than confirming the original data. That gave us an opportunity to check out the general validity of the first set of coding and the capacity to compare the structure and performance of these systems before and after the WECS/IIMI intervention (see Lam and Shivakoti, 2002), as we do in depth in Chapters 3 and 4.

3 Processes and procedures of an innovative development intervention initiated in 1985 in the Middle Hills of Nepal

In 1985, the International Irrigation Management Institute (IIMI) entered into agreement with the Water and Energy Commission Secretariat (WECS) of Nepal, as briefly described in the last chapter, to undertake action-research on sustainable interventions in irrigation systems. In 1986, IIMI, which was based in Sri Lanka, established one of its country offices in Nepal with the mandate to promote the study of farmer-managed irrigation systems (FMIS). While WECS was not an implementing agency, one of its mandates was to undertake action-research on the different aspects of the water sector, including the irrigation sector. In order to ensure that the farmers of the action-research area would benefit directly from the research initiative, the Ford Foundation provided two kinds of grants: one was given to IIMI to support action-research on public intervention in FMIS and the other was for action-research on actual implementation of the intervention program.

OBJECTIVES OF THE ACTION-RESEARCH PROJECT

Multiple objectives have been pursued in the action-research project in Indrawati River basin. One objective was to establish low-cost procedures for identifying the relative needs of all systems in an area, allowing the selection of systems for assistance where the greatest impact on food production could be made. Another objective was to develop and test the methods for delivering assistance that would enhance the farmers' management capability for the operation and maintenance of irrigation systems while the physical infrastructure was being improved.

The goal of expanding the existing FMIS included assurance that they remain farmer-managed systems. It was assumed that such an approach required the full participation of the farmers in the identification of the available resources and limitations. Furthermore, it was anticipated that farmers' participation in the improvement activities under the guidance of competent technicians would give the farmers experience in maintaining the physical system and expose them to the management skills essential for mobilizing local resources.

PROJECT AREA

In collaboration with WECS, the location for action-research on public intervention in FMIS was selected. The project site selected was in the Middle Hills region of Nepal, specifically, the Upper Indrawati River basin in Sindhupalchok District (see Figure 3.1). Proximity to Kathmandu for supervision of the research was a major consideration in site selection. The project staff could travel from Kathmandu to the Indrawati River in about an hour and a half. It then took anywhere from one to three hours to travel on foot to the irrigation systems located in the hills on either side of the Indrawati River.

At the lower end of the project area where the elevation is about 1000 m, three irrigated crops can be grown each year. At the higher elevations, low temperatures limit the growing season to two crops. Rice is the main irrigated crop in the monsoon season. If the water supply is adequate, an irrigated spring rice crop is also grown in the hot, dry season preceding the monsoon. If water is limited, maize may be grown instead of rice before the monsoon. Wheat and potatoes are the predominant irrigated winter crops.

To allow systematic identification of existing systems, the river basin hydrologic boundaries were used to define the project area. This reduced travel time and simplified supervision since it is the basin's drainage pattern that determines the location of the systems, not political boundaries (WECS/IIMI, 1990).

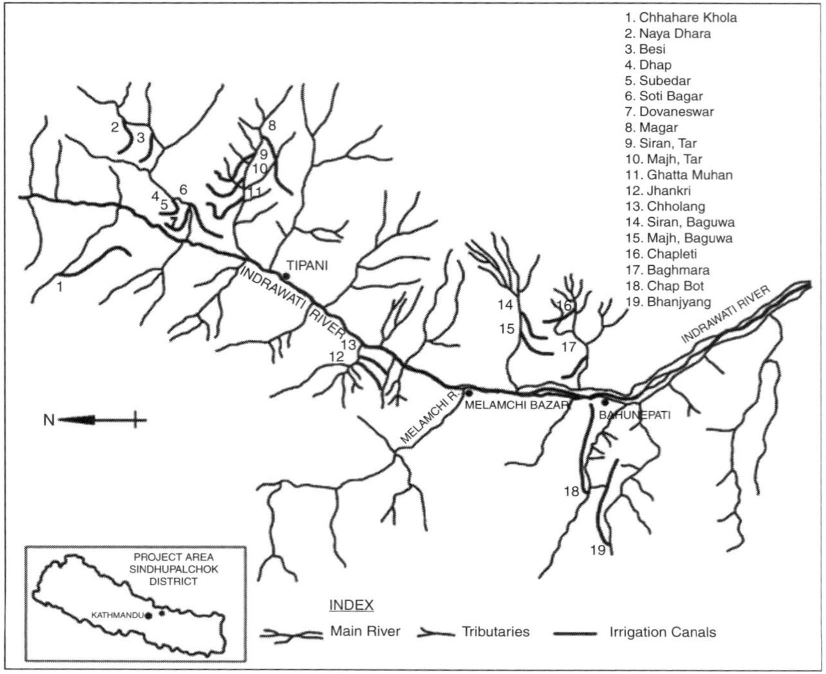
1. Chhahare Khola
2. Naya Dhara
3. Besi
4. Dhap
5. Subedar
6. Soti Bagar
7. Dovaneswar
8. Magar
9. Siran, Tar
10. Majh, Tar
11. Ghatta Muhan
12. Jhankri
13. Chholang
14. Siran, Baguwa
15. Majh, Baguwa
16. Chapleti
17. Baghmara
18. Chap Bot
19. Bhanjyang
TIPANI
INDRAWATI RIVER
MELAMCHI R.
MELAMCHI BAZAR
BAHUNEPATI
INDRAWATI RIVER
N
PROJECT AREA
SINDHUPALCHOK
DISTRICT
KATHMANDU
INDEX
Main River
Tributaries
Irrigation Canals

Source: Authors.

Figure 3.1 Project area sketch map of Indrawati watershed, Sindhupalchok, showing names and locations of the 19 systems that received assistance

THE PROCEDURES USED BY THE ACTION-RESEARCH PROJECT

The WECS/IIMI and Ford Foundation action-research project carried out the following procedures in the Sindhupalchok District in order to provide assistance to FMIS. These procedures may provide inspiration for other programs to adapt as appropriate to their needs. This is a hill area where the Indrawati River has cut deep into the valley, making the water from this large snow-fed river nearly inaccessible to farmers for irrigation. To develop irrigation, farmers have constructed diversions on the small high-gradient tributary streams to the Indrawati River. These streams have destructive floods in the monsoon and only a small spring-fed discharge in the dry season. Farmers have built contour canals, often across rock cliffs and through unstable slopes, to irrigate terrace fields.

System Identification and Selection

The initial activity was planned into two stages. In the first stage, the objective was to identify all irrigation systems in the 200 sq km project area. In the second stage, a reconnaissance/inventory preparation study was undertaken in order to determine the location and resource base of each system within the project area. On the basis of the inventory, potential candidates for assistance were identified. Different criteria were used to identify potential candidate systems: potential for expansion of command area, intensification of crops, or reducing maintenance. The reconnaissance/inventory study identified 119 systems with canals longer than 0.5 km and command areas larger than 5 ha. Twenty-two of these systems were identified as candidates for improvement on the basis of having the possibility of command area expansion and extra water resources available.

After identifying the potential 22 systems, a rapid appraisal was carried out in order to identify basic information about the water availability and characteristics of the irrigation users' organization to manage operation and maintenance. During the rapid appraisal process, information was also collected about the need for physical improvement. After the stage-one and stage-two activities, 19 systems out of 119 were selected for assistance (see Figure 3.1).

Table 3.1   Basic information about the irrigation systems

Name of system	Location	Year of establishment
Chhahare Khola	Baruwa VDC[b]–8	1971
Naya Dhara	Thangpal Kot VDC–1, 2, 3, 4	1973
Besi	Thangpaldhap VDC–1	1946
Dhap and Subedar[a]	Thangpaldhap VDC	1897
Soti Bagar	Thangpaldhap VDC	1974
Dovaneswar	Thangpaldhap VDC–3 & 4	1979
Magar	Bhote Namlang VDC	1895
Siran, Tar	Thangpaldhap VDC–7	1974
Majh, Tar	Thangpaldhap VDC–7	1974
Ghatta Muhan	Thangpaldhap VDC	1960–61
Jhankri	Dubachaur VDC–9	1802
Chholang	Dubachaur VDC–8 & 9	1895
Siran, Baguwa	Sikharpur VDC	1980
Majh, Baguwa	Sikharpur VDC–5, 6, 9	1965
Chapleti	Sikharpur VDC–9	1973
Baghmara	Sikharpur VDC–9	1960
Chap Bot	Bansbari VDC–2	1969
Bhanjyang	Bhanjyang VDC–4	200 years ago

Notes:
a.   Dhap Kulo and Subedar Kulo (numbers 4 and 5 in Figure 3.1), with two intakes from the same water source, serve the overlapping command area, so they are put in one row. However, the total number of systems receiving assistance in the WECS/IIMI program was 19.
b.   VDC = Village Development Committee.

Source: Field Survey (1990).

Implementation of Improvement of Selected Irrigation Systems

Based on the results of the rapid appraisal of these systems, the first dialogue between the farmers and agency personnel took place in the selected system (see Table 3.1). The purpose of the first dialogue was to obtain information on the number of beneficiaries, role and strength of the beneficiary organization, and irrigation management practices. Critical areas that needed physical improvement were also identified. The first dialogue was important because it established a rapport between the farmers and technical and social groups from the project side.

In addition to collecting the basic information, the technical team with the participation of the farmers also collected relevant data for the design of the new irrigation structure. The farmers were then asked whether they wanted to participate in such an assistance program and what contributions they could make. After the farmers had agreed to the terms and conditions of assistance and once each system was notified, and the total amount of money available for improvement determined, a second dialogue with the farmers took place.

In the second dialogue, the following activities occurred. First, a tentative list of irrigation improvements was prepared. In establishing the priority of the various physical improvements, the farmers were asked to help rank all of the desired physical improvements into three groups according to priority:

1. The highest priority was placed on improvements necessary for expansion of the system but difficult for farmers without assistance.
2. The second priority was assigned to work that would improve system operation and maintenance.
3. The third-priority improvements included work that farmers could accomplish using their own skills, labor, and materials (Yoder, 1991a, p. 57).

The farmers, in consultation with the technicians, also needed to take into consideration the budget ceiling set by the government's financial contribution in deciding what improvement work would be undertaken using government assistance and what they would do on their own. In most cases, the farmers could usually accomplish the earthworks on their own.

Second, the farmers were informed of the full amount of money allocated to be spent on their system and that all of those funds were to be spent on their system. If they could save money on first-priority work, they would be able to use it for second- and even third-priority work. The intention was to create a positive incentive for the farmers to use the project funds with great care.

Third, a users' organization was formed to be responsible for several tasks, including (1) the identification of existing and future water users (from the expanded area) and the land area each irrigated; (2) preparation and acceptance by all water users of a plan for water allocation to the new area; (3) preparation of a plan, including rules, for supervising the improvements to be made and for future management of operation and maintenance; and (4) setting the requirements and rates for free and paid labor mobilization (WECS/ IIMI, 1990, p. 20).

Fourth, the farmers and the engineers designed the structures for the improvement of the irrigation system. During this time, some shifting of priorities and changes in design took place. Fifth, assistance to FMIS involved both physical and managerial improvement of the system. For example, to help farmers strengthen their management capacity, a social organizer was present to help them conduct regular meetings. In addition, members of an irrigators' executive committee, who were elected by the farmers, were trained in recording the minutes of meetings, keeping records on labor mobilization, and keeping financial accounts.

Furthermore, farmers could also strengthen their managerial and organizational capacity by visiting other similar systems to learn from their experiences. A farmer-to-farmer training program was organized for members of the irrigation organization so they could observe improved irrigation practices in another system and learn from the farmers in that system. Several farmers selected by the farmers themselves participated in the farmer-to-farmer training program. After their return from the visit to other systems, the participant farmers organized meetings with the users of the system and explained what they had observed in other systems and what might be adopted, and what needed modification in the context of their own system. Thus, a larger number of farmers were exposed to new and improved irrigation management systems during the planning phase.

MANAGEMENT IMPROVEMENTS THROUGH FARMER-TO-FARMER TRAINING

A major problem identified during rapid appraisal, according to WECS/IIMI, was that 'the water users of the systems selected for assistance did not function as organized bodies to manage the operation and maintenance activities of their canals' (1990, p. 18). Thus, during the rapid appraisal study, farmer training for irrigation management in each system was identified as a priority in implementation of the project. Although the field supervisors assisted farmer management by advising committee members in group decisions, keeping records and mobilization of labor, the result was not satisfactory. Members of the project decided, therefore, to try a series of farmer-to-farmer training tours as a method of extending ideas about effective governance and management of irrigation systems.

The purpose of the farmer-to-farmer training program, according to Naresh Pradhan, was 'to stimulate the transfer of experience from farmers in well-managed systems to those in poorly-managed systems through site visits, informal exchanges, and guided discussions' (1987, p. 1). The project organized farmer-to-farmer training for five groups of farmers from these 19 irrigation systems with each group consisting of 15 farmers. Each group was accompanied by two facilitators – one of whom was a member of a host system and the other a research assistant who was hired by WECS/IIMI for the project period. The host farmers from the well-managed systems also worked as consultants. These consultant-farmers inspected the canals and structures of the systems and discussed the similarities and differences in their own systems and made suggestions for improvements.

During the tour, the trainee farmers were taken to the intake and canal of the host system guided by a group of host farmers. Though the timing of the tour had been arranged to coincide with the annual meeting of the canal's organization, the trainee farmers were taken first to meet the host system's committee members. These host farmers described the ways they had devised to deal with issues such as labor mobilization for emergency maintenance, water allocation and distribution, conflict management, and the structure of the organization. The facilitator usually raised questions that covered important issues. In the general meetings, however, the visiting farmers only observed the procedures of a general meeting and did not participate in it.

The trainee-farmers were also exposed to the host system's constitution, minutes, and records of the labor contributed by the farmers. By the end of the second day, the visiting farmers had a better idea about management problems in their own systems. The farmers were taken to more than one system; and during the second successful system visit, the farmers started comparing their own systems with the two systems observed and discussed their problems with the host farmers.

Upon return from the visit to the two systems, the farmers who participated in the farmer-to-farmer training program held a meeting of the farmer users of their own system and described what they had observed, and what they thought could or could not be done. Hence, larger groups of the farmers were oriented about the features of the better-managed irrigation systems. The observations became the reference point for innovations in the water user associations and the formulation of rules and regulation for better irrigation management. This is reflected by the farewell advice of the host farmer as follows:

> You farmers have described your irrigation systems as having an illness. Now you realize that you have the medicine for the illness in your pocket. Other members of your systems still don't recognize that there is medicine available. You must step forward and explain that unless you all take the medicine your system will not improve. It may be bitter medicine to take but after your system operates effectively all will be happy that you have taken the medicine. (Yoder, 1991b, p. 11)

In summary, during the process of the first and second dialogues, and also during the physical and management improvement period, field supervision was carried out by teams that consisted of engineers, overseers, agriculturists, social scientists, and persons with construction skills. The construction activities were to be a 'training exercise for the users' organizations in making decisions, establishing rules, managing conflicts, mobilizing labor and keeping records' (WECS/IIMI, 1990, p. 20).

During the third dialogue, the farmers' contributions and the role and responsibilities of the technicians were defined. The implementation of physical improvement also began. During this third-dialogue phase, the farmers requested many additional changes in design as they better understood the actual dimensions and other characteristics of the structures that were to be built. Designs were modified to accommodate site-specific characteristics as excavation and construction work progressed. One out of the three consulting groups took charge of the improvement of six out of the 19 systems selected for assistance. It was reported that out of 47 structures that were initially designed for the six systems, 30 were modified to meet the farmers' requests or to better fit the site conditions during construction. Eight of the 47 initial structures were dropped by the farmers during construction in favor of adding 42 others totaling the same cost but better fitting their priorities. In essence, the project's commitment to full farmer participation and farmer acceptance of the designs required that the technicians resolve the farmers' dissatisfaction over any aspect of the project (Bhattarai, 1990).

Specifically, technical assistance from three engineering companies was given to the irrigation systems for their improvement. One team included engineers, hydrologists, sociologists and supervisors who were responsible for gathering information for the project. The responsibility of another team was to mobilize the farmers by helping them to form a functioning Water User Association (WUA) where such organization did not exist. The management committee of the WUA was required to supervise the construction activities and take the responsibility for operation and management of the system after improvement. The third team was charged with the task of interacting and motivating the farmers.

As a result of such effort, the farmers agreed to undertake all necessary earthwork by arranging labor contributions and to conduct all other work, including the transport of construction materials, at a rate lower than both the village labor rate and the district rate. The savings thus made by using a lower rate were used on the additional system-improvement activities beyond the first-priority work for which the funding was allocated. Hence, the intervention in Indrawati River basin was like the three-legged table with the balanced approach of (1) the government, (2) the active participation of the beneficiary farmers, and (3) the consulting companies performing both technical and social mobilization. All contributed to the effectiveness of the program (Acharya, 1990).

INITIAL RESULTS FROM THE ACTION-RESEARCH PROJECT OF INNOVATIVE INTERVENTION

The following results initially emerged from the action-research project. First, the existing accounting procedures and audit systems needed to be remodeled to allow greater flexibility. Structures needed to be adapted to the requirements of the actual field site and local needs. Some of these changes could not be anticipated until actual work began. A flexible design procedure was essential so that the structures could be adjusted as necessary.

Second, farmers found it difficult to understand blueprints and design sketches. Engineers needed to use three-dimensional scaled models to explain to the farmers how the proposed structures looked and operated. Observing actual structures in operation in another farmer-managed system helped. Almost 40 percent of the structures required design changes during the construction phase. Many of the structures that were originally designed were unnecessary. The engineers had used blueprints and design sketches to explain the construction and use of the structures to the farmers. However, it was not until the farmers actually saw the structures begin to take shape, and where they would be installed, that they fully comprehended the blueprints. Then, they requested changes.

Third, it was important to maintain transparent accounts – accounts of income and expenditures open to all for inspection. The beneficiaries needed to be given information on the status of the project. Being informed allowed the beneficiaries to determine how best to mobilize their own resources and assume management responsibilities. This created a strong bond among users through enhanced ownership of the irrigation system.

Fourth, bureaucratic reorientation was essential in implementing effective assistance to FMIS. It was necessary to make a distinction between rehabilitation and assistance to FMIS. Rehabilitation of the system referred to making improvements to the physical condition of the system. Assistance to FMIS referred to helping the farmers to improve their system. Hence, assistance was provided to improve managerial capacity as well as physical capacity.

It was also important that irrigation officials change their orientation from physical rehabilitation *only* to assistance. Assistance was successful when the farmers were allowed to participate fully in all activities. The reorientation needed to occur at the central level as well as the field level. The change in attitude is best achieved when officials can observe the results of the implementation of this type of assistance program.

In addition to these four points, the following results of the innovative intervention were reported in the report prepared by IIMI and WECS (WECS/IIMI, 1990).

Competitive Construction Cost

Table 3.2 shows that assistance to the 19 systems allowed expansion of the irrigated area commanded by the canals by more than

Table 3.2   Irrigable area and cost of improvements to 19 farmer-managed systems

Name of system	Existing command area (ha)	Command area expansion (ha)	Total irrigable area (ha)	Project grant (NRs)	Cost per irrigable hectare (NRs)
Chhahare Khola	126	37	163	126615	777
Naya Dhara	55	55	110	139720	1270
Besi	65	20	85	119839	1410
Dhap and Subedar[a]	30	35	65	85000	1308
Soti Bagar	19	11	30	150699	5023
Dovaneswar	2	10	12	74807	6234
Magar	100	43	143	160805	1125
Siran, Tar	18	6	24	136789	5700
Majh, Tar	71	16	87	114321	1314
Ghatta Muhan	23	10	33	124321	3767
Jhankri	18	13	31	91707	2958
Chholang	23	14	37	116066	3137
Siran, Baguwa	18	19	37	57488	1554
Majh, Baguwa	13	20	33	113541	3441
Chapleti	8	15	23	78065	3394
Baghmara	3	6	9	44433	4937
Chap Bot	12	5	17	71630	4214
Bhanjyang	21	14	35	65178	1862
Total	625	349	974	1871024	
Average cost per irrigable hectare					1921
Consultant and WECS supervision support				1192747	
Tools supplied				82182	
Farmer training				55000	
Average cost of supervision per irrigable hectare					1365
Total cost of improvement per irrigable hectare					3286

Note: a.   Numbers 4 and 5 from Figure 3.1 are put in one row. Altogether, there are 19 systems that received WECS/IIMI assistance.

Source: Adapted from WECS/IIMI (1990, p. 29).

50 percent. The expenses incurred were recorded separately by each system and the record book was open for inspection by all users, the consultant and WECS and IIMI staff. The cost based on the grant to each system was just under NRs (Nepalese rupees) 2000/ha (about NRs 22 = US$ 1 at the time the grant was received). With supervision included, the cost of physical and management improvements was about NRs 3300 (US$ 150) per hectare. This is in the same cost range as other agencies that have provided assistance[1] to farmer systems in the hills using participatory methods such as the Farm Irrigation and Water Utilization Division (FIWUD) averaging NRs 3400/ha, and the Agricultural Development Bank of Nepal (ADB/N), which cost about NRs 4600/ha. Although the cost of supervision was high, the close participation enhanced the productivity of the money spent.

More important than the low capital cost per hectare of the grant was the effect of intensive supervision and farmer training tours in motivating farmers to use the grant resource productively and to augment it with their own labor. This resulted in nearly all of the improvements identified by the farmers and consultant (including second- and third-priority work) being completed even though the budget was expected to cover only the improvements of first priority. Table 3.3 shows that farmer involvement in construction resulted in a 38 percent contribution from the farmers; about half of the systems managed substantial labor mobilization from their own resources.

Averaged over all the systems, farmer participation can be credited with increasing the value of the grant by about 140 percent, where the volume of work completed is at the rates given in the national norms for rate analysis. Most of the increases in value of the work done can be credited to the efficiency of work accomplished by farmer participation over what would have been required if contractors had been used.

Although a great deal of time and effort was required to bring about effective farmer participation, and the project got off to a slow start with delays for design modifications, ultimately it resulted in an extraordinary farmer response during construction. Once farmers were convinced that they were getting what they needed from the project, they worked hard to get the most out of it.

Positive Agriculture Results

Assistance for physical improvements was completed just before the monsoon rice season in 1989. No time was available for most

Table 3.3   *Savings in cost of improvements due to farmer*
participation (amount in NRs '000)

Name of system	First-priority work		Saving	Farmers' contri-bution	Work com-pleted	Effective increase d/a
	(a)	(b)		(c)	(d)	(e)
	Grant	Actual expendi-ture	(a–b)/a (%)	NRs '000	NRs '000	(%)
Chhahare Khola	127	62	51	3	168	132
Naya Dhara*	140	—	—	21	245	175
Besi*	120	—	—	10	221	184
Dhap and Subedar**	85	35	59	4	154	181
Soti Bagar	151	83	45	1	167	111
Dovaneswar	75	68	9	1	89	119
Magar	161	133	17	1	192	119
Siran, Tar	137	40	71	1	214	156
Majh, Tar	114	96	16	1	143	125
Ghatta Muhan	124	82	34	0	170	137
Jhankri	92	28	70	1	108	117
Chholang	116	41	65	1	136	117
Siran, Baguwa	57	42	26	25	81	142
Majh, Baguwa	114	85	25	42	170	149
Chapleti	78	60	5	19	109	140
Baghmara	44	30	32	12	73	166
Chap Bot	72	60	17	16	86	119
Bhanjyang	65	50	23	15	102	157
Total	1872	995	38	174	2628	140

Notes:
a. Grant amount allocated to the systems to complete most first-priority work as estimated using national norms.
b. Grant money expenditure for completing first-priority work – money saved (a–b) was used for second- and third-priority work.
c. All unpaid labor (calculated as the number of person-days of labor multiplied by the district wage rate) plus the difference between the district rate and a lower wage rate as agreed to by farmers in some systems to reduce cost.
d. Value of work completed as computed using national norms. This is higher than (a + b) because: (1) estimates computed by norms are generally high and (2) work efficiency due to farmer participation was very high.

Table 3.3   (continued)

---

e.   Effectiveness of the farmer participation in accomplishing more than estimated by the national norms.
*   Naya Dhara and Besi Kulo systems are not included because information on the actual cost is not available.
**   Numbers 4 and 5 from Figure 3.1 are put in one row. Altogether, there are 19 systems that received WECS/IIMI assistance.

Source: Adapted from WECS/IIMI (1990, p. 30).

farmers to convert their upland fields into level terraces for growing rice. Farmers in one system reported that on the few hectares they were able to terrace, production shifted from an average of 1.7 tons per hectare (t/ha) of millet to nearly 3.0 t/ha of rice. Farmers indicated it would take them four or five years to complete the terrace building, but wide-scale work was underway.

Seventy-six percent of the farmers interviewed after the monsoon rice harvest indicated they had previously grown rain-fed rice on land that they were able to irrigate for the first time after the system was improved. The total sample of 16 ha that shifted from rain-fed to irrigated rice reported an average increase in yield of about 50 percent, from 1.5 to 2.2 t/ha. A sample of 106 farmers with more than 44 ha of rice land that had intermittent access to irrigation in the past reported that on average, yields went from 1.2 to 2.3 t/ha, or an increase of about 90 percent. Many farmers indicated that their yields that year were reduced due to a severe hailstorm, but that they expected to get a much higher return in future years. In this first cropping season, farmer practices regarding fertilizer did not change. All the increase in production was due to improved irrigation. As reliability is established, farmers will use fertilizer and other inputs resulting in even higher impact. Active agricultural extension could shorten the time required to achieve full production (WECS/IIMI, 1990).

A more intensive evaluation was undertaken by WECS in 1990 to determine the total impact on agricultural production. Even in the earlier stage, there were clear indications that rapid change was taking place.

ANALYZING THE EFFECTS OF INTERVENTION IN SINDHUPALCHOK

A survey of different productivity indicators, as well as area coverage, shows the increase of winter crop area before and after intervention (see Table 3 in WECS/IIMI, 1990, p. 34). The coverage of potato crop increased from 5.9 ha to 28.6 ha, oilseed from 16.8 to 48.3 ha, wheat from 108.5 to 170 ha and vegetables from 3.7 to 13.5 ha. The increase in the area coverage of crops and of crop productivity after intervention shows that the intervention process was largely successful; improvements in the indicators are significant. While increases in values of productivity indicators are obvious, some interesting questions remain unanswered.

First, even though there was an increase in productivity, can we draw the conclusion that such an increase was caused by the intervention effort? Is it possible that the increase is only a result of chance? Second, what is the magnitude of different factors that affect productivity? Specifically, how substantial is the effect of the intervention process on productivity? Third, how does the effect of intervention operate to affect productivity? Did the intervention process bring about a one-shot effect and cause an abrupt increase in productivity? Or did the intervention effort manifest its impact on productivity by mediating the effects of other input factors, such as the number of labor days spent on the maintenance of the canals? These questions relate to a broader issue concerning government assistance to FMIS: how should the intervention proceed?

There have been two major perspectives of what government intervention means and, hence, how it should operate so as to achieve a high level of effectiveness. The first perceives government intervention as a process of transferring resources to farmers by government agencies. The assumption of this perspective is that if only farmers are given adequate resources, irrigation water will flow automatically. An intervention process is seen as a one-shot process. Thus, the major concern of an intervention process is to make sure that the magnitude of the shot is strong enough.

The second perspective posits that intervention is more likely to be effective when it enhances farmers' ability to manage their systems. Through intervention, farmers are enabled to mobilize themselves better to maintain the resources and to engage in self-governing activities concerning appropriation and maintenance. From this perspective, both the direct and mediating effects of an intervention process are important.

NOTE

1. Assistance here implies provision of minor inputs and farmer involvement in making the improvements. The Department of Irrigation tends to use a rehabilitation approach where major input, usually through contractors, is given to improve a system to standards and with methods that would be applied to building a new system.

4 Evaluating an innovative development intervention a decade and a half later

ANALYZING THE LONG-TERM PERFORMANCE OF AN INTERVENTION

In the third chapter, we reported on and described an ingenious intervention program designed and initiated in 1985 jointly by the Water and Energy Commission Secretariat (WECS) of Nepal and the International Irrigation Management Institute (IIMI) for 19 irrigation systems located in the Sindhupalchok District of Nepal (Yoder, 1986, 1994; P. Pradhan, 1989a, 1989b). Instead of following the then established 'best practices' that focused on spending large funds for engineering works and imposing a top-down planning process, the project extensively involved farmers in deciding what should be done, and built in learning mechanisms that allowed farmers to adapt to the changing environment. In this chapter, we will report on our efforts to evaluate the performance of this innovative effort over time using both conventional statistical analysis as well as Qualitative Comparative Analysis (QCA) (Ragin, 1987).

The purpose of this evaluation is not to try to provide another piece of evidence to show that engineering works alone do not work. That is well established (see Chapter 1). Prior studies of engineering solutions, however, have tended to be either in-depth historical analysis of a small number of irrigation systems or statistical analysis of variables measured at one time for a larger number of systems. While these studies offer important insights, their static, descriptive orientation has not enabled their authors to decipher the mechanisms through which improved engineering infrastructure impacts on irrigation performance and unfolds across time depending on other causal processes. Instead of looking at engineering solutions as competing with other factors to affect irrigation performance, we focus on how improved engineering works interact with other factors in affecting irrigation performance, and to identify conditions under which improved engineering works can make a difference in irrigation performance. Policy analysts have tended to rely on analysis of interventions viewed as simple additive processes rather than complex configural processes. Our efforts to measure the long-term performance of the intervention program are intended to decipher the complex, over-time processes initiated by an innovative development project that tried to avoid imposing a 'best-practices' cure-all on local farmers.

INITIAL EVALUATION OF IMPACT

In our initial effort to understand the impact of this intervention, we entered the data collected by the WECS/IIMI team in the Nepal Irrigation Institutions and Systems (NIIS) database that we described in previous chapters. In our NIIS database, we were already coding variables related to various physical and institutional features as well as irrigation performance of the systems we were able to study in the field (Lam, 1996a, 1998). By augmenting the data collected by the WECS/IIMI team in 1985 about the systems before the intervention, this 'Time Slice 1' serves as a benchmark against which the impact of the intervention can be assessed.

In 1991, members of the NIIS research team visited these same 19 systems to conduct a second round of data collection using the coding instruments developed for the NIIS database. The collected information describes the action situations of the systems a few years after the intervention, which constitutes the 'Time Slice 2' data. Two of the authors of this book conducted statistical analysis to compare the performance at Time Slice 1 and Time Slice 2 (Lam and Shivakoti, 2002). Statistical models were specified to capture two approaches to understanding intervention efforts. The first approach looks at an intervention as a one-shot process of transferring resources to farmers. As long as the 'shot' is strong enough and farmers are given adequate resources, irrigation will be improved. The second approach emphasizes the facilitative role of intervention: intervention affects performance through enabling farmers to better utilize the physical, human, and social resources that are available to them. Both the direct and mediating effects of an intervention process are important.

The findings of the initial statistical analysis clearly indicated that intervention is not a one-shot additive process of transferring resources to the farmers. Specifying the effect as both direct and mediating is better able to capture how intervention impinges upon performance. Intervention has an essential role to play in determining the relationships between input variables, such as maintenance effort, and agricultural productivity as measured by cropping intensity at the tail end. The analysis corroborates the argument that intervention should 'enhance' rather than 'replace' the efforts of local farmers in irrigation management (Shrestha, 1988; Shivakoti, 1992).

THE SUSTAINABILITY OF THE EFFECT OF THE INTERVENTION

While the initial evaluation suggested that the intervention brought about significant improvements in technical efficiency and agricultural productivity in the first few years after the completion of the intervention, interesting questions remained as to whether the improvements were only temporary, and whether the enhancing effect of the intervention has been sustained. If there is variation in the sustainability of intervention effects among the systems, what are the factors that might have contributed to such variation? These questions are of major interest to researchers and international donors, who have seen ample examples of intervention improvements beginning to dissipate soon after the completion of a project.

To assess the long-term sustainability of the intervention effect of the WECS/IIMI project, the NIIS team visited the 19 systems again in 1999 (eight years after the second visit). Using the same coding forms, the team collected information on the physical and social aspects of the systems, as well as performance measures.[1] The information collected constitutes the core of the 'Time Slice 3' data reported on herein. To supplement the NIIS data and to capture the processes of evolution and change of the systems, the NIIS team visited the systems again in 2001 to conduct a series of intensive qualitative interviews. In-depth interviews were conducted, focusing on the processes of change in performance and institutional arrangements. In particular, farmers were asked to identify major events and disturbances since the intervention, and to discuss how the disturbances impinged upon the evolution of rules and collective action. We sought to capture the temporal dimension of farmers' adaptation to change, and to understand whether and how the intervention effect has affected the adaptation process.[2]

Patterns of Change of Irrigation Performance

The availability of structured information about the 19 systems in three time slices plus in-depth information about diverse processes allows us to study how irrigation performance has changed over the years. Particularly, the data provide useful information on both short-term and longer-term effects of the intervention, which allow for an analysis of temporal patterns of performance. We measured the short-term effect by the change in performance from Time Slice 1 to Time Slice 2, and the longer-term effect by the change from Time Slice 2 to Time Slice 3. The summation of the two effects gives the net effect of intervention. In particular, we examine five key measures of irrigation performance: size of irrigated areas, technical efficiency of irrigation infrastructure, water adequacy, tail-end cropping intensity, and levels of deprivation in a system. The patterns of change of these performance measures can shed light on the way the effects of the intervention have unfolded over time.

Size of irrigated area

Table 4.1 provides information on the size of irrigated area of the 19 systems in the three time slices; the average sizes in the three time slices are 37.1 ha, 52.58 ha, and 58.84 ha, respectively. A comparison of these figures shows that the WECS/IIMI intervention initially brought about substantial increases in the size of irrigated area of the systems and that this expansion leveled off after the initial increase.

A note of caution is warranted. While the intervention succeeded in expanding the size of the systems in general, the magnitude of the expansion varied among the systems. In fact, an expansion did not occur in every system. The pattern of change of the size of irrigated area of the systems over time shows diversity. During the period between Time Slices 1 and 2, 17 of the 19 systems showed an expansion of irrigated area; one system shrank in size, and one system had no change. In the period between Time Slices 2 and 3, while ten of

Table 4.1   The size of irrigated areas of the irrigation systems

Name of system	Size of irrigated area over time (hectares)			Impact of intervention
	T1 (1985)	T2 (1991)	T3 (1999)	
Chhahare Khola	126	163	151	Expanded–fluctuating
Naya Dhara	55	110	110	Expanded
Besi	65	85	85	Expanded
Dhap	70	50	60	Deteriorated–fluctuating
Subedar	40	40	60	Expanded
Soti Bagar	19	30	32	Expanded
Dovaneswar	2	12	6	Expanded–fluctuating
Magar	100	143	140	Expanded–fluctuating
Siran, Tar	18	24	48	Expanded
Majh, Tar	71	87	71	No change–fluctuating
Ghatta Muhan	23	33	35	Expanded
Jhankri	18	31	31	Expanded
Chholang	23	37	37	Expanded
Siran, Baguwa	18	37	50	Expanded
Majh, Baguwa	13	33	65	Expanded
Chapleti	8	23	15	Expanded–fluctuating
Baghmara	3	9	12	Expanded
Chap Bot	12	17	18	Expanded
Bhanjyang	21	35	40	Expanded

Note: Information about the size of the irrigated area of Dhap and Subedar, which used to overlap command areas, was provided by the farmers during fieldwork for the NIIS data collection.

Source: Lam and Ostrom (2010, p. 8).

the systems had an increase in the irrigated area, five systems showed a decrease, and four remained unchanged. Both the short-term and longer-term effects of the intervention on the size of irrigated area are statistically insignificant at the 0.1 level. Overall, the intervention has succeeded in increasing the size of irrigated area in a majority of systems, and the drive for expansion of irrigated areas has persisted, although the magnitude of expansion has leveled off in the longer run.

An interesting pattern is that six of the 19 systems experienced a reversal of the intervention effect. Five of these systems started with a positive effect between Time Slices 1 and 2, followed by a negative effect between Time Slices 2 and 3; the remaining system had a negative short-term effect, then a positive longer-term effect. Thirteen of the 19 systems showed a positive effect all the way. Overall, the intervention has succeeded in increasing the size of irrigated area in a majority of systems, and the drive for expansion of irrigated areas has persisted, although the magnitude of expansion has leveled off after the initial expansion.

Technical effectiveness of irrigation infrastructure
Improving technical efficiency of irrigation infrastructure was another major objective of the WECS/IIMI project. Technical efficiency concerns whether the physical infrastructure is able to deliver water so that farmers are able to obtain as high crop yields as is feasible, given the other constraints they face. In the NIIS database, technical efficiency is measured by a four-point scale. The coded values of technical efficiency in the three time slices are: 4 = highly effective; 3 = moderately effective; 2 = moderately ineffective; and 1 = highly ineffective. Table 4.2 provides information about the levels of technical efficiency of the systems in the three time slices.

The short-term effect of the intervention is obvious. Of the 19 systems, 14 showed a short-term improvement in technical efficiency, while five showed no impact. Given that engineering improvement work was a major element of the intervention, such an outcome should not be surprising. Although the improvement works involved only primitive technologies, such as putting in gabion boxes to strengthen the water diversion devices and providing simple canal lining, they helped make water flow more predictable and also minimize seepage. With better control of the temporal and spatial availability of water, a higher level of technical efficiency can be attained.

The longer-term effect, however, shows an interesting pattern. Of the 19 systems, all, except for two that showed no impact, experienced a deterioration of technical efficiency in the period between Time Slices 2 and 3. As a result, 13 of the 19 systems have gone through a reversal of the intervention effect – they experienced an improvement in technical efficiency in the initial years after the intervention, which was then followed by a deterioration of efficiency in the longer term. The differences between Time Slices 1 and 2, and between Time Slices 2 and 3, are statistically significant at the 0.1 level. In terms of the net effect, only two of the 13 reversing systems

Table 4.2   Technical efficiency of irrigation infrastructure

Name of system	Technical efficiency[a]			Impact of intervention
	T1 (1985)	T2 (1991)	T3 (1999)	
Chhahare Khola	2	4	2	No change–fluctuating
Naya Dhara	3	4	3	No change–fluctuating
Besi	3	3	2	Deteriorated
Dhap	2	4	3	Improved–fluctuating
Subedar	2	4	2	No change–fluctuating
Soti Bagar	2	4	3	Improved–fluctuating
Dovaneswar	3	4	3	No change–fluctuating
Magar	3	3	3	No change
Siran, Tar	3	3	2	Deteriorated
Majh, Ta	3	4	2	Deteriorated–fluctuating
Ghatta Muhan	2	4	2	No change–fluctuating
Jhankri	2	4	2	No change–fluctuating
Chholang	3	4	2	Deteriorated–fluctuating
Siran, Baguwa	3	3	2	Deteriorated
Majh, Baguwa	3	4	3	No change–fluctuating
Chapleti	1	3	3	Improved
Baghmara	4	4	2	Deteriorated
Chap Bot	2	4	2	No change–fluctuating
Bhanjyang	3	4	2	Deteriorated–fluctuating

Note: a.   Technical efficiency coded as (1) highly ineffective, (2) moderately ineffective, (3) moderately effective, (4) highly effective.

Source: Lam and Ostrom (2010, p. 9).

have gained net improvement in technical efficiency; three systems in fact wound up having a level of technical efficiency even lower than before the intervention. The six systems that did not experience a reversal had a neutral initial response to intervention in the short run, and then experienced deterioration in the longer run. Chapleti (system 16) is the only system that had a positive initial effect and was able to keep the improvement.

Water adequacy
While engineers might be interested in achieving a high level of technical efficiency, farmers are more concerned about whether those of

Table 4.3a   Average water adequacy for 19 systems

	T1 (1985)	T2 (1991)	T3 (1999)
Spring tail end	2.36	1.54	2.0
Winter tail end	2.06	1.26	1.25
Monsoon tail end	1.64	1.0	1.0
Spring head end	2.0	1.37	1.43
Winter head end	1.8	1.2	1.1
Monsoon head end	1.53	1.0	1.0

Note: Water adequacy as coded for individual systems: (1) abundance, (2) limited, (3) scarce, (4) non-existence.

Source: Lam and Ostrom (2010, p. 11).

them who want water can actually get it. That a system has a high level of technical efficiency is no guarantee that farmers actually receive water when they need it. It is not uncommon that water is efficiently delivered to the wrong place at the wrong time. On the other hand, a system that has a relatively low level of technical efficiency could provide adequate water to its farmers if the water delivery process is managed properly (Chambers, 1988; Lam, 1998). Water adequacy is determined not only by the physical setting and engineering infrastructure but also the effectiveness of water management as well as the level and timing of demand for water (Gill, 1991; Burns, 1993).

The average levels of water adequacy at the head end and the tail end across all systems in three different seasons are shown in Table 4.3a. The short-term effect of the intervention on water adequacy is obvious. In Time Slice 2, an average system has a level of water adequacy higher than 2 in a four-point scale, with 1 being the highest,[3] at both the head end and the tail end in all seasons. The data also suggest that the positive effect of the intervention on water adequacy has persisted in most systems. Only a slight deterioration is found among systems at the tail end in spring; yet the difference is not statistically significant at the 0.1 level.

Comparing the pattern of change of water adequacy with that of technical efficiency, one can see interesting relationships between intervention, technical efficiency, and water adequacy. Intervention brought about an improvement in technical efficiency through enhancing infrastructure works, which in turn fostered an improvement in water adequacy. As time passed, the infrastructure improvement works wore out; the technical efficiency in almost all systems started to deteriorate. The decrease in technical efficiency, however, was not necessarily accompanied by deterioration of water adequacy. Of the 16 systems for which information on water adequacy at the tail end in winter is available, only two of them experienced deterioration in water adequacy in the period between Time Slices 2 and 3 (see Table 4.3b) and 12 of them showed consistently improved water adequacy. This suggests that improved infrastructure may help to launch farmers' collective action for irrigation management; after the launch, however, as long as a good management order can be maintained, farmers may be able to attain a high level of water adequacy even with less well-constructed or maintained infrastructure.

Tail-end cropping intensity

A common measure of the agricultural productivity of an irrigation system is the cropping intensity at the tail end of the system. One crop per year on a plot of land equals 100 percent; two crops mean 200 percent. If only part of the land is covered with two crops during the year, the cropping intensity will be less than 200 percent. Table 4.4 shows the tail-end cropping intensity of the systems in the three time slices. The data suggest that the short-term effect of the intervention on tail-end intensity was mixed. Only six of the 18 systems for which we have complete data had an increase in agricultural productivity between Time Slices 1 and 2, six systems had a decrease, and six showed no effect.[4]

The pattern of change of cropping intensity varied across the individual systems. Ten of the 19 systems recorded an improvement in the period between Time Slices 2 and 3. Only five systems experienced deterioration in the same period; two systems showed no effect. In terms of the net effect (comparing the cropping intensities in Time Slices 1 and 3), nine systems showed a net increase, although the magnitude of the increase in two of the systems was relatively minor. Only four systems showed continual improvement in both time slices. Four systems experienced a net decrease in cropping intensity; one of them experienced a neutral short-term effect and another a neutral longer-term effect. Only two systems showed a continual decrease in agricultural productivity.

Several patterns could be identified on the basis of the analysis of the tail-end intensity data. First, the effect of intervention on

Table 4.3b   Water adequacy at the tail end in winter

Name of system	Tail-end water adequacy[a]			Impact of intervention
	T1 (1985)	T2 (1991)	T3 (1999)	
Chhahare Khola	2	2	1	Improved
Naya Dhara	2	1	1	Improved
Besi	2	1	1	Improved
Dhap	2	1	1	Improved
Subedar	2	2	Missing data	Unknown
Soti Bagar	2	1	1	Improved
Dovaneswar	Missing data	1	1	Unknown
Magar	3	1	2	Improved–fluctuating
Siran, Tar	2	1	Missing data	Unknown
Majh, Tar	2	1	1	Improved
Ghatta Muhan	Missing data	1	Missing data	Unknown
Jhankri	2	1	1	Improved
Chholang	2	2	1	Improved
Siran, Baguwa	2	2	1	Improved
Majh, Baguwa	2	2	1	Improved
Chapleti	4	1	1	Improved
Baghmara	1	1	4	Deteriorated
Chap Bot	1	1	1	No change
Bhanjyang	2	1	1	Improved

Note: a.   Water adequacy coded as: (1) abundance, (2) limited, (3) scarce, (4) non-existence.

Source: Lam and Ostrom (2010, p. 11).

tail-end intensity takes time to factor in. Second, most of the systems recorded a positive impact in the longer term, although the magnitude of the improvement varies substantially. Third, considerable diversity can be found in the trajectory of change in tail-end intensity among the systems; many systems have experienced a reversal of effect in the course of development. Obviously, the relationship between intervention and agricultural productivity is not as simple

Table 4.4   Tail-end cropping intensities

Name of system	Tail-end intensity			Impact of intervention
	T1 (1985)	T2 (1991)	T3 (1999)	
Chhahare Khola	200	192	200	No change–fluctuating*
Naya Dhara	200	200	205	Improved*
Besi	200	235	210	Improved–fluctuating*
Dhap	250	250	290	Improved
Subedar	250	270	300	Improved
Soti Bagar	150	215	300	Improved
Dovaneswar	300	200	300	No change–fluctuating
Magar	190	200	210	Improved*
Siran, Tar	255	250	Missing data	Unknown
Majh, Tar	300	230	230	Deteriorated
Ghatta Muhan	271	270	Missing data	Unknown
Jhankri	200	270	300	Improved
Chholang	220	220	300	Improved
Siran, Baguwa	300	285	250	Deteriorated
Majh, Baguwa	280	300	300	Improved*
Chapleti	Missing data	300	200	Unknown
Baghmara	300	300	100	Deteriorated
Chap Bot	270	270	300	Improved
Bhanjyang	260	220	115	Deteriorated

Note: *Small magnitude of change.

Source: Lam and Ostrom (2010, p. 12).

and straightforward as sometimes thought. Agricultural productivity tends to be affected by a complex array of factors; whether and how an intervention can bring an improvement often depends on how it configures with the other factors (Lam and Shivakoti, 2002).

Level of deprivation
A major measure of deprivation is whether some irrigators in the system are consistently disadvantaged in the allocation of water. Table 4.5 shows the number of irrigation systems in which some

Table 4.5   Levels of deprivation

Are there appropri-ators who are consistently disadvantaged in irrigation?	Time slice		
	1 (1985)	2 (1991)	3 (1999)
Yes	8	5	0
	(50%)	(26.32%)	(0.00%)
No	8	14	17
	(50%)	(73.68%)	(100%)
Total	16	19	17
	(100%)	(100%)	(100%)

Note: Chi Square = 11.02; *P*-value = 0.004.

Source: Lam and Ostrom (2010, p. 13).

irrigators are consistently disadvantaged in the three time slices. Before the intervention (i.e., in Time Slice 1), half of the systems had the problem of deprivation. The number dropped in Time Slice 2, when only one-fourth of the systems had the problem of deprivation. This positive trend continued. In Time Slice 3, *none* of the systems had the problem of deprivation.

Making Sense of the Changing Irrigation Performance

Analysis of the changing patterns of irrigation performance of the 19 systems sheds light on how the intervention has affected irrigation performance. First, the effects of intervention on the technical aspects of irrigation management are conspicuous in the short run. Due to improved infrastructure, the size of irrigated area increased and the technical efficiency of the irrigation systems also improved. Yet these positive effects dissipated, or leveled off, in the longer run. In particular, the analysis suggests that the improvement of technical efficiency has withered away in almost all of the systems.

Second, while better technical efficiency might have helped improve water adequacy in the short run, it does not explain the persistence of adequate water supplies in the longer run. The analysis suggests that, even after the technical efficiency leveled off, farmers in many systems were able to maintain a relatively high level of water adequacy. Third, generally speaking, the intervention brought about consistent improvement in the way that farmers interact with one another. Deprivation has decreased and more equitable water allocation has been attained. Fourth, while one might expect that improved irrigation infrastructure and management might help bring about better agricultural productivity, the analysis portrays a more complicated picture. Much fluctuation can be found in the patterns of change of tail-end cropping intensity among the systems, suggesting that a variety of factors interplay to affect agricultural productivity instead of a single cause.

While the patterns of relationships among the various perform-ance measures provide interesting hints on how the intervention has affected irrigation performance, building a statistical model that explicates and explains the intervention's effect on irriga-tion performance across time is a substantial challenge. Our data describe only three snapshots of an unfolding process. Studying these snapshots per se might not fully capture the dynamics of the process. Other than data problems, we are also faced with serious methodological problems of using conventional statistical analysis, which is variable-oriented, to capture the unfolding effect of an intervention. The effects of individual variables in conventional sta-tistical analysis are assumed to be independent from each other. The estimate of a variable is calculated by averaging out the effect of the variable across all observations (Ragin, 1987). Obviously, a major problem of this assumption is that the effect of one variable is often contingent on the value of other variables.

Often, the combinatorial effects of several variables affect outcomes. The combinatorial effects are usually not linearly related.[5] Also, sta-tistical analysis by its variable-oriented nature cannot adequately deal with the time dimension of social phenomena. Some causes and effects, such as those involved in examining the effect of intervention, take a long time for them to unfold and to be discernable (Pierson, 2003, 2004). Notwithstanding the notorious modeling problem of accurately measuring dynamics in statistical models, it is also difficult to specify the complex dynamics of institutional change in a statistical model.

Given these methodological problems, in-depth qualitative inter-views are an important complementary research methodology to capture the changes and unfolding events in the systems. The field-work conducted in 2001 by the NIIS team, as mentioned earlier, was devoted to asking farmers in-depth questions about the governance and operation of these systems over time. In response to the checklist of questions, farmers responded with narratives of major events that have occurred since the intervention, and of the way these events have unfolded and impinged upon irrigation management. To synthesize the in-depth responses, we will apply Boolean algebra, which emphasizes the holistic nature of explanation of social phenomena. Instead of treating an empirical case as a mere collection of values related to some variables, the Boolean method is case-oriented, treating a case as a configuration of causal and outcome conditions. The basis of explanation is not the correlation of variables but a systematic comparison of configurations of causes that produce a similar outcome. 'Rather than focus on the net effects of causal conditions, case-oriented explanations emphasize their combined effects' (Ragin and Sonnett, 2005, p. 1). Unlike conventional small-N qualitative case study methods, however, the Boolean method extends the holistic analysis to the studies of a larger number of cases, and allows a more systematic distinction between necessary and sufficient causal conditions (Rihoux, 2008).

COPING WITH COMPLEXITY AND CHANGE

The information provided by farmers in the qualitative interviews suggests that the 19 irrigation systems in the Indrawati watershed took different paths of development after the completion of the WECS/IIMI intervention. Some of them were able to build upon the improved social and physical endowments brought about by the intervention and thrived; some were only able to reap the initial benefits of the intervention and failed to sustain effective irrigation management; others simply failed to make a good use of the opportunity and have remained poor performers.

An interesting question of major policy importance is whether one could identify a set of causal conditions, amidst diverse experiences, that are conducive to the persistence of the intervention effect. Based upon a review of the literature as well as the qualitative interviews conducted with farmers from the 19 systems, we identify five factors that appear to explain why some systems have continued at a higher level of performance and why there are differences in the longer-term effects of the intervention project. Specifically, we will examine how (1) continual assistance to improve infrastructure, (2) the existence of written rules, (3) the imposition of fines and punishment, (4) leadership and (5) collective action may have affected the sustainability of the intervention effect and hence irrigation performance. We will now apply the Boolean analysis to examine how these causal conditions configure to bring about particular patterns of outcome.

Continual Infrastructure Investment

Since the completion of the WECS/IIMI intervention project, a number of systems have received further external assistance mainly to fix or further improve the lining and diversion structures of the systems. The amount of funds varied, ranging from NRs 10 000 to NRs 2 300 000.[6] The sources of funding included a local Rural Development Committee, District Irrigation Offices, as well as international donors such as UNDP. For those who think that irrigation management mainly concerns moving water from where it is available to where it is needed, continual investment in irrigation infrastructure is obviously an important factor affecting irrigation performance. Presumably, continual infrastructure investment can help consolidate the improved physical endowment, hence making the intervention effect persistent.

The data analysis presented above, however, challenges this view. The level of technical efficiency in almost all of the systems dropped in the period between Time Slices 2 and 3. Interestingly, of the two systems that did not experience a drop in technical efficiency in the period (Magar [system 8] and Chapleti [system 16]), one did not receive any further assistance, and the other received continued assistance worth merely NRs 10 000, the smallest amount among all those assisted. Thus, the evidence is that continual infrastructure investment, in general, did not help maintain, let alone improve, technical efficiency. Yet, interesting questions remain as to whether the systems that have received continuing assistance are more likely to sustain a higher level of performance as compared with those that have not; and how continual infrastructure investment might configure with other factors to affect irrigation performance and the sustainability of the intervention effect.

Written Rules

A major element of the WECS/IIMI intervention project was that farmers in a system were required to develop their own set of rules for irrigation management after the project was completed. The case materials show that the WECS/IIMI project did trigger rule-crafting efforts in most systems. The scope and the comprehensiveness of the rules, however, varied across the systems. While the rules developed for most systems mainly concerned water allocation and canal maintenance, farmers in some systems worked out extensive rules dealing with issues such as arranging payments from farmers to support full-time watchmen. In particular, systems in which water management served multiple purposes, including generating hydropower and grain mills, tended to develop a more sophisticated set of rules with a broader scope. In Naya Dhara (system 2), for example, a set of rules was developed for water management for both energy generation and irrigation. A caveat is warranted, however. The existence of rules does not necessarily mean that the rules are strictly followed. In some systems, written rules existed but were often not followed because of a lack of commitment and enforcement. In other systems, strong leaders were able to coordinate farmers' action that rendered strict application of formal rules unnecessary.

In some of these systems, in which farmers did not draft a set of written rules, farmers just failed to work with one another to engage in rule-crafting activities. Yet, systems existed in which farmers were able to organize collective action for water distribution and maintenance on the basis of verbal agreements and understanding. In these systems, the norms and common understanding that had evolved over the years already provided a good basis for collective action. Farmers did not see the need to turn these norms into written rules. Particularly in systems in which there were strong leaders, the leadership already provided the necessary focal point for collective action.

Fines

The third condition that may affect irrigation management is whether farmers have made efforts to punish those who violate the rules or free-ride on the efforts of others. Only nine of the 19 systems had worked out rules imposing fines. A review of the experiences of the system reveals several interesting patterns. First, systems that had a set of written rules for system operation and maintenance did not necessarily also have a provision for imposing fines. In many systems in which leadership was strong and a good working order was in place, farmers did not see the need for imposing fines. Second, whether and how a fine provision has affected farmers' incentives to contribute to irrigation operation and maintenance depends on the configuration of factors that constitute farmers' action situations.

In Dhap (system 4), for example, where the Water User Committee (WUC) provided an arena for farmers to resolve conflicts effectively, imposing fines was considered unnecessary. In Chhahare Khola (system 1), on the other hand, although a small group of local leaders was able to maintain a certain level of collective action, the lack of a provision for fining violations was considered to be a major reason why many farmers did not participate in canal repairs and maintenance. Third, some farmers did not think of the imposition of fines as a useful means for enhancing collective action. As a farmer in Majh Baguwa (system 15) succinctly put it, 'only the honest and sincere people pay the fines.' In fact, farmers in most systems were hesitant to impose fines even if there were provisions in the rules for their imposition.

Consistent Leadership

A major factor that may affect the long-term viability of irrigation management is the presence of effective leadership that is able to adapt to change. When the WECS/IIMI project was implemented, the designers were well aware that building up strong leadership in the irrigation systems was important for the project's success. Systems that were chosen for the project were in general free from major conflicts that could prevent the building of effective leadership. In particular, the Time Slice 1 data in the NIIS database show that, although a certain degree of ethnic heterogeneity existed in all the 19 systems, it did not affect effective communication among farmers in any of them.[7] As discussed above, one of the major elements of the project was to enhance the organizing abilities of local leaders through training so that they could serve as the catalysts for collective action in the local community. As a result, in almost all systems, a WUC, or the equivalent of it, was set up to provide leadership.

Local leaders performed important functions. In most systems, water allocation and system maintenance were coordinated by the WUC. In Ghatta Muhan (system 11), for example, the chairperson was a very powerful leader who decided on the order of water distribution for the other farmers in the system to follow. In Jhankri (system 12), canal cleansing and system maintenance were mainly coordinated by the chairman of the WUC. Moreover, local leaders often served as the arbitrator for resolving conflicts among farmers, or between farmers and outsiders.

Yet, the existence of a WUC does not necessarily mean that leadership exists. Leadership, in the context of irrigation management in Nepal, is embedded in the broader social relationships in the local community. In a few systems in which existing social capital might not have been as solid, the WECS/IIMI project seems to have had limited success in building up leadership. What happened in Baghmara (system 17) is a case in point. Over the years, ethnic issues have escalated to affect the solidity of the community. Although the WECS/IIMI project helped farmers to set up a WUC, the committee became inactive shortly after the completion of the project; farmers reported that the committee has not held meetings ever since.

For systems that had strong leadership at the beginning of the project, how to cope with leadership change so as to maintain consistent leadership poses a challenge. What happened in Magar (system 8) is illustrative. When the WECS/IIMI project was implemented, a WUC was formed under the leadership of Mr. Batan Singh Tamang. Mr. Tamang was instrumental not only in the process of implementing the intervention project but also in sustaining farmers' collective action in managing irrigation operation and maintenance. His leadership was so effective that farmers did not see the need to put the rules in written form. The death of Mr. Tamang in 1998 presented a big shock to the system. While there were some other potential leaders, farmers failed to agree on a successor. As a result, the irrigation working order unraveled rapidly. What happened in Magar is not unique. In other systems, such as Chap Bot (system 18) and Dovaneswar (system 7), the demise of old leaders also caused confusion and bewilderment. In fact, the general situation is that the more frequently the leadership has changed, the less effective the WUC has become. Systems that have been better able to maintain a good working order and good performance are often characterized by the existence of a core group of active leaders who provide a level of consistent leadership.

Co-management existed in three systems. In the 1990s, hydropower-generating facilities were constructed on Dhap Kulo and Naya Dhara Kulo. As water was no longer only used for irrigation, an integrated management framework was put in place. A water resource committee was set up to replace the existing WUA;

the new committee was given the authority to oversee both the energy and irrigation matters. Expanding the scope of water management brought in both opportunities and constraints. On one hand, bringing two groups of users (irrigation and energy) under the auspices of one institutional framework inevitably increased the likelihood of conflicts. While the farmers would like to have the flow of water under control, the energy users would prefer continual flow of a large volume of water in the canals. Farmers reported that the conflict between farmers at the head end and the energy users was particularly severe. On the other hand, bringing in the energy users also brought about a form of checks and balances. The energy users had the incentive to see to it that the operation and maintenance of the water system was in good shape, and that a set of rules for water delivery and canal repair be developed and followed. More importantly, leaders of the energy sector have been proven to be a valuable asset for water management; farmers commented that, since the new institutional framework was put in place, leadership has improved, which in turn has brought about improvement in agricultural productivity.

Co-management could take another form, however. In Besi Kulo, the WUA was abolished in the 1990s, and the responsibility for irrigation management was shifted to the Ward, a local general-purpose political-administrative unit. The Ward chairman was given the responsibility of 'supervising' the operation of the system. As one would expect, the chairman did not spend too much time and effort on irrigation management. A result was a rapid unraveling of management for irrigation. Since the institutional change, no meetings have been held to discuss irrigation management. The management of the system was, in the farmers' words, 'on an individual basis.' Ironically, although a set of written rules for irrigation operation and maintenance existed in the system, most farmers were not even aware of its existence.

Collective Action

The existence of written rules and the presence of active leaders are not likely to bring about good performance if farmers are not willing to be involved in collective action for system operation and maintenance. Collective action helps build trust, norms, and common understanding among farmers in the process of working with one another (Poteete et al., 2010). Improvement in trust and common understanding can, in turn, serve as the foundation for collective problem-solving in other contingencies. The success of collective action hinges upon a variety of factors. In Magar, for instance, the lack of leadership resulted in the unraveling of collective action among farmers and brought about low levels of technical efficiency of the system. In Majh Tar (system 10), however, farmers were able to maintain a certain level of collective action despite a lack of leadership in the system.

Maintaining a certain level of collective action among farmers is always a challenge (Dietz et al., 2003). A seemingly minor event could trigger the unraveling of collective action easily. What happened in Siran, Baguwa (system 14) is a good case in point. When the WECS/IIMI project was first implemented, farmers in the system were enthusiastic about the endeavor and were willing to engage in collective action for irrigation management. As farmers recollected, it was the time when 'an environment of trust' prevailed, which enabled them to attain a high level of technical efficiency of their system. The effective working order began to fall apart in this system, however, when some farmers at the head end stopped participating in collective maintenance works. They argued that a continual flow of water in the canals could adversely affect their lands. It was unfortunate that the leadership in the system was also going through changes at the same time. The new leaders failed to resolve the conflict between the farmers at the head end and their neighbors at the middle and the tail ends. The situation deteriorated rapidly. Conflicts have become a frequent occurrence, particularly during the dry season from March to June when water is scarce. As the farmers put it, a 'crisis of trust' is reducing the technical efficiency and agricultural performance of the system.

CONFIGURATIONS OF CAUSAL CONDITIONS THAT SUSTAIN INTERVENTION EFFECTS

As argued above, the five causal conditions do not operate independently to affect irrigation performance in a linear manner. To understand how these causal conditions combined to affect the sustainability of intervention effects, it is necessary to identify possible configurations that lead to particular outcomes, and to specify the necessary and sufficient causal conditions. In this section, we will employ the Boolean method to help compare the configurations of causal factors leading to the sustainable intervention effect identified in the 19 systems.

All causal conditions, or variables, in the Boolean analysis are dichotomized.[8] A condition is coded as being either present (PRESENT) or absent (ABSENT). Such simplification helps focus the analysis on the structure of relationships among causal conditions rather than on the competition among variables to explain the outcome. In our analysis, we want to know what configurations of causal conditions account for the persistence of intervention effect. The outcome, or the dependent variable, is whether persistent improvement in irrigation performance existed in a system or not. In this analysis, we will examine two measures of irrigation performance, namely, the adequacy of water at the tail end of the system in winter season, and agricultural productivity at the tail end. We focus on performance measures at the tail end, because irrigation at the tail end faces a more challenging task environment, especially in winter, which has the least amount of rainfall. Hence, these two measures should be most sensitive to changes in the causal conditions.

The two outcome variables are meant to capture not only whether there was an improvement in irrigation performance 15 years after the intervention in 1985, but also whether the improvement was persistent. The coding for 'water adequacy at the tail end in winter season' (W) was based on the information provided in Table 4.3b. We coded a system as PRESENT if the net intervention effect on water adequacy at the tail end in winter was positive, *and* there was no reversal or fluctuation in intervention effect. For all other situations, we coded them as ABSENT. For example, the intervention brought about a short-term positive impact on water adequacy in Magar (system 8), followed by a negative impact. Although the net effect was positive, we coded the system as ABSENT. To provide a simplified notation, we adopted the convention of using the upper-case letter (W) to indicate a PRESENT value and the lowercase letter (w) to denote an ABSENT value (see Table 4.6).

The coding for the 'cropping intensity at the tail end' (T) was based on information provided in Table 4.4. Again, the coding was intended to capture not only whether there was a net improvement in tail-end cropping intensity across time but also whether the improvement had been consistent. Unlike water adequacy, which is mainly

Table 4.6   Notations used in coding and Boolean analysis

Dependent variable	Symbol	PRESENT	ABSENT
Persistent improvement in water adequacy at the tail end in winter	*W*	W	w
Persistent and significant increase in tail-end cropping intensity	*T*	T	t
Causal condition			
Continual assistance on infrastructure improvement	*A*	A	a
The existence of a set of formal rules for irrigation operation and maintenance	*R*	R	r
The existence of provisions of fines	*F*	F	f
The existence of consistent leadership	*L*	L	l
The existence of collective action among farmers for system maintenance	*C*	C	c

Source: Lam and Ostrom (2010, p. 19).

determined by irrigation management, tail-end cropping intensity is affected by an array of factors other than whether the irrigation system is well maintained. As a result, a small change in the tail-end cropping intensity might reflect more the effect of other contextual factors than the effect of the intervention. So in our coding, we imposed a higher standard for what we mean by an improvement. If a system had a persistent increase in tail-end intensity but the net increase was less than a quarter of a crop, we coded it as ABSENT. These cases are indicated with an '*' in Table 4.4.

There are thus five causal conditions in the analysis, namely whether a system has received further infrastructure assistance since the completion of the WECS/IIMI project (*A*), whether farmers in a system have been able to develop a set of written rules for irrigation operation and maintenance (*R*), whether farmers have worked out provisions for imposing fines (*F*), whether the leadership in a system has been able to maintain continuity and to adapt to the changing environment (*L*), and whether farmers have been able to maintain collective action in system maintenance (*C*). By reviewing what happened in the systems a decade and a half after the intervention, we sought to identify major events and their impacts on irrigation management, and also to understand the processes and dynamics of change. As mentioned above, the causal conditions are coded as either PRESENT or ABSENT. Again, the uppercase letter denotes a PRESENT value and a lowercase letter an ABSENT value for the systems for which we have relevant data (see Tables 4.7 and 4.8). For some causal conditions, such as *A* and *R*, the coding is rather straightforward. But for some other conditions such as *L* and *C*, delineating a PRESENT value from an ABSENT value requires careful interpretation of the local history of the systems, as well as an exercise of judgment. Fortunately, in the qualitative interviews, farmers did offer vivid descriptions of events and their comments, which have provided a good basis for our coding.

The Boolean analysis starts with constructing a Truth Table, which lays out all the configurations of the five causal conditions that exist among the 19 cases. The Truth Table for water adequacy at the tail end in winter (*W*) is shown in Table 4.7 for all 15 systems for which we have complete data. Eight configurations generate W, two configurations generate w, and one configuration generates contradictory outcomes. Note that there are only 11 unique configurations for these 15 systems in the Truth Table because several configurations predict water adequacy for more than one system (as shown in the last two columns of Table 4.7). For this study, we are primarily interested in configurations generating W, so only the configurations that generate W will be used for subsequent analysis.

A problem exists, however, of how to handle the configuration that generates contradictory outcomes. Practically, the contradiction might suggest that there were some other causal conditions affecting the outcome that we did not take into account in our analysis; or that it is simply a result of randomness. The Boolean convention in dealing with contradictions is to adopt a threshold value for deciding whether the particular configuration is more likely to generate W or w. We adopted the 50 percent threshold, meaning that unless a particular configuration generates more W than w, we will not consider it as a configuration generating W. Accordingly, the configuration that generated contradictory outcomes failed to pass the threshold and was not included in our analysis. In Table 4.7, the configurations used for analysis are shaded.

The fsQCA software was used to minimize the configurations generating W and to devise an equation that succinctly describes the relationships between W and the configurations of causal conditions:

Table 4.7 *Truth Table for persistent improvement of water adequacy at tail end in winter*

Five causal conditions					Number of systems	
A	R	F	L	C	W	w
ABSENT	PRESENT	ABSENT	PRESENT	PRESENT	1	1
ABSENT	PRESENT	PRESENT	PRESENT	PRESENT	2	0
PRESENT	PRESENT	ABSENT	PRESENT	PRESENT	2	0
PRESENT	PRESENT	PRESENT	PRESENT	PRESENT	2	0
ABSENT	ABSENT	ABSENT	ABSENT	PRESENT	1	0
ABSENT	PRESENT	ABSENT	ABSENT	PRESENT	1	0
ABSENT	ABSENT	PRESENT	PRESENT	PRESENT	0	1
PRESENT	ABSENT	ABSENT	ABSENT	ABSENT	0	1
PRESENT	PRESENT	PRESENT	ABSENT	ABSENT	1	0
PRESENT	PRESENT	ABSENT	ABSENT	PRESENT	1	0
PRESENT	PRESENT	PRESENT	ABSENT	PRESENT	1	0

Source: Lam and Ostrom (2010, p. 20).

$$W = ACR + aICf + AIRF + LCRF \qquad (4.1)$$

Two questions are of major policy interest. First, how important is continual infrastructure investment for sustaining adequate water supply in the systems? Is it a necessary condition? If not, what are the contexts for it to be present for water adequacy to occur? Second, it has often been argued that experiences of collective action are the basis for the accumulation of social capital. One of the major objectives of the WECS/IIMI intervention, in fact, was to encourage and enable farmers to participate more fully in irrigation management. Exactly how does collective action relate to water adequacy? What are the supporting conditions with which collective action can bring about water adequacy?

To address these questions, we rearrange the Equation (4.1) to:[9]

$$W = AR(C + IF) + CLRF + Calf \qquad (4.2)$$

In Boolean algebra, addition is equivalent to the logical operator OR; multiplication means the conjunction of causal conditions. Equation (4.2) is composed of three groups of configurations, meaning that if any one of three groups of configurations is obtained, there is persistent improvement of water adequacy. The three groups are configurations where, respectively, continual infrastructure investment must be present (A), may be present or absent (i.e., *A* is irrelevant), and must be absent (a).

The first group of configurations suggests that continual infrastructure investment can bring about persistent improvement in water adequacy *only if* farmers have been able to develop a set of written rules for system operation and maintenance (the presence of *A must* come with the presence of *R*). This suggests that the debate about whether physical infrastructure or social infrastructure is more important is misplaced; neither of them works without the other. The existence of these two factors alone, however, is necessary but not sufficient; it has to be in a context where either farmers are able to engage in collective action, or fines are imposed for rule violations with a lack of strong leadership. The policy implication of such a finding is that intervention projects might help bring in infrastructural investment (A) and put in place formal rules (R) in an irrigation system. These two factors, however, are not sufficient by themselves to bring about an effective outcome. Either they have to be complemented by a certain degree of collective action among farmers based upon common understanding and norms, or, in systems in which farmers' voluntary collective action does not exist, they have to be backed up by strict implementation of fines without a strong leader dominating the management of the systems.

The second group of configurations in Equation (4.2) concerns situations in which continual infrastructure investment is irrelevant to irrigation performance. In systems in which there are written rules, consistent leadership, a certain degree of farmers' collective action and also strict implementation of fines, one can find persistent improvement in tail-end water adequacy in winter no matter whether there has been infrastructure investment or not. This pattern suggests that, to make up for the positive impact of continual infrastructure investment, *all* the other factors have to be present to sustain the positive impact of the intervention. An implication is that there is a limit as to the positive impact of continual infrastructure investment that can be substituted. The third group of configurations refers to situations in which a system has not received any infrastructure assistance since the completion of the WECS/IIMI project. Without continual assistance, collective action of farmers becomes a very important factor affecting irrigation performance. Moreover, the analysis suggests that the collective action is effective only if it is *not* organized on the basis of punishment.

Now we turn to the analysis of configurations of causal conditions leading to persistent increase in cropping intensity at the tail end. Again, the analysis starts with constructing a Truth Table as shown in Table 4.8. Two configurations generate T, seven configurations generate t, and two configurations generate contradictory outcomes. Again, the configurations that generate contradictory outcomes were dealt with by the 50 percent threshold. Only one of the two configurations that generate contradictory outcomes was included in our analysis. In Table 4.8, the shaded configurations are those we include in the analysis.

The fsQCA software was used to generate an equation that lays out the relationships between T and the configurations of causal conditions:

$$T = aCRf + LCRf \qquad (4.3)$$

We rearranged the equation to:[10]

Table 4.8 Truth Table for persistent increase in cropping intensity at tail end

A	R	F	L	C	Number of systems	
		Five causal conditions			T	t
PRESENT	PRESENT	ABSENT	PRESENT	PRESENT	2	1
ABSENT	PRESENT	ABSENT	PRESENT	PRESENT	2	0
ABSENT	PRESENT	PRESENT	PRESENT	PRESENT	0	2
PRESENT	PRESENT	PRESENT	PRESENT	PRESENT	1	1
ABSENT	PRESENT	ABSENT	ABSENT	PRESENT	1	0
ABSENT	ABSENT	ABSENT	PRESENT	PRESENT	0	1
ABSENT	ABSENT	PRESENT	PRESENT	PRESENT	0	1
PRESENT	ABSENT	ABSENT	ABSENT	ABSENT	0	1
PRESENT	PRESENT	PRESENT	ABSENT	ABSENT	0	1
PRESENT	PRESENT	ABSENT	ABSENT	PRESENT	0	1
PRESENT	PRESENT	PRESENT	ABSENT	PRESENT	0	1

Source: Lam and Ostrom (2010, p. 20).

$$T = CRf (a + L) \qquad (4.4)$$

Equation (4.4) provides a succinct statement about the relationships between the configurations of the causal conditions and the existence of persistent increase in tail-end cropping intensity. The term CRf is the necessary element of configurations for persistent increase in tail-end cropping intensity. It means that the simultaneous existence of collective action, written rules, and the absence of a provision of fines are the necessary conditions without which sustained improvement in agricultural productivity would be impossible. Such a pattern is consistent with the findings of prior research that effective irrigation management hinges upon a good working order sustained by farmers' continued involvement and a set of rules (Ostrom, 1990, 1992; Lam, 1998). With collective action and the rules in place, formal punishment would not be necessary, or it could even be harmful to collective action.

The necessary elements per se are insufficient. They can bring about sustained improvement in agricultural productivity only if either one of two additional conditions exists – the presence of consistent leadership and the absence of continual external assistance for infrastructure improvement. Obviously, these two additional conditions are consistent with and complementary to the necessary elements. As we have found in the qualitative interviews, local leaders have played an important role in enhancing and maintaining farmers' collective action in the project irrigation systems. Not only have they provided a locus for coordinating collective action but they also served as an arbitrator in resolving conflicts and disputes among farmers. In fact, leadership is particularly important in the context where farmers tend to be hesitant in resorting to formal punishment, and consider discussion and arbitration as a better means for conflict resolution.

LESSONS LEARNED

What are the lessons we have learned from this innovative project? The lessons are general and cannot be picked up and applied routinely in other settings without knowledge of these settings; for example, incessant rainfalls may make continued assistance necessary. We hope that our findings provide insights into how it is possible to help farmers help themselves to maintain better irrigation facilities and greater agricultural outcomes without massive infusion of funds.

While developing robust local institutions to support the operation and maintenance of engineering infrastructure should not be viewed as a panacea, it needs to be part of the design of projects intended to have a long-term, positive impact on a high proportion of systems that receive external assistance. Further, the designers of projects can learn from, as well as contribute to, the knowledge base of local farmers. When farmers have no voice in the design of systems that are supposed to help them, we can expect few successes over time.

The WECS/IIMI intervention project to assist 19 farmer-managed irrigation systems located in the Indrawati watershed in Nepal was designed with a view to developing and testing methods for delivering assistance that could enhance farmers' organizing ability for irrigation operation and maintenance at the same time as the irrigation infrastructure was improved. In this chapter, we have drawn upon multiple rounds of data collection for the systems involved in the project to assess and understand how the intervention has affected the operation and performance of the systems in a decade and a half after completion.

Before one asks the question of what can be done to help the farmers improve irrigation performance, one has to appreciate the challenges and complexity involved in managing irrigation in a region where the natural environment is hostile and the material condition is austere. Torrential rains, flooding, and landslides are common occurrences in the monsoon season, which often damage the primitive irrigation infrastructure that farmers have struggled to construct, and render system maintenance extremely difficult and costly. It is not surprising that, in many instances, when government officials and donors came in to try to 'assist' the farmers, fixing the engineering infrastructure was often the first thing that came to mind.

Our analysis of the experience of the WECS/IIMI intervention, however, has shown that infrastructure fixes can improve technical efficiency only in the short run. In most of the cases, the improvement in technical efficiency as the result of intervention withered away soon after the completion of the intervention. Such a situation should come as no surprise. For farmers who engage in constant struggle with the tough environment, working together to fix and rebuild their systems is simply part of effective irrigation management. In fact, in some irrigation systems in Nepal, the diversion structure is built of primitive materials intentionally so that during the monsoon season water could be stopped from getting into the system to flood the canals and farmers' fields (Lam, 1998). Given the challenging environment, to maintain a high level of technical efficiency by continual infrastructure investment is not, and should not be, a realistic objective. In fact, our analysis has shown that continual infrastructure assistance cannot sustain a high level of technical efficiency.

Does it mean that infrastructure fixes are irrelevant to efforts to help farmers improve irrigation management and performance? The answer is no. Our analysis has suggested that, in most of the 19 systems involved in the WECS/IIMI project, the improved technical efficiency of irrigation infrastructure did bring about an improvement in water adequacy, which has persisted even after the improved technical efficiency withered away. As discussed above, the WECS/IIMI intervention was designed to involve farmers in the processes of planning and implementing the infrastructure works to the greatest possible extent. The infrastructure improvement works provided not only incentives for farmers, who could see for themselves how their effort could make a difference, but also effective opportunities for farmers to develop working relationships with one another. As long as a good working order can be maintained, a high level of water adequacy can be achieved.

Maintaining a good working order, of course, is no less a challenge than coping with the capricious physical environment. It requires a mastery of human artisanship – the abilities and skills required for working with one another for mutual betterment. A major focus of the WECS/IIMI intervention was to help farmers improve such abilities and skills and to avoid swamping them with expensive works that might make them dependent on external aid. Through farmer-to-farmer training, getting the farmers involved in project implementation, identifying local leaders and helping farmers to work out rules, the intervention set the momentum for farmers' self-organization. The project avoided the 'best-practices trap' of relying primarily on infrastructure improvement designed entirely by external experts.

Given their different history and social-political backgrounds prior to the intervention, the 19 systems have taken on different paths for self-organization. Some have been able to build upon the momentum and thrived; others have failed to sustain the physical improvements achieved early in the process. It would be naive to think there is a single recipe for developing human artisanship. Yet our analysis has suggested that as long as farmers are willing to maintain a certain level of collective action, and a core of local entrepreneurs exists to provide leadership and adjustments to changes, it is possible for the farmers to build on the momentum introduced by the intervention to attain consistently high levels of performance over time.

NOTES

1. During the initial period of project development and our first revisit in 1991, the Maoist insurgency had not yet erupted in Nepal. During the later visits, the region was divided: those villages near the road, which was under the control of the Nepali army, were relatively peaceful. The systems located in remote areas were under Maoist control and were also relatively peaceful. The villages served by one of the systems in the middle – Majh Baguwa – did face considerable challenge as villagers were divided in their loyalty and faced army patrols during the day and Maoist patrols at night. While very disruptive of social relationships since the villagers were themselves divided, this disruption did not adversely affect the overall performance of this system due in part to its setting and the relative ease of obtaining water and needing only a minimum level of repair and maintenance. Only modest 'out migration' from this rural setting occurred during the time of our study, and thus few external remittances were introduced.

2. The qualitative interviews aimed at capturing the process of institutional change as well as major developments that had occurred in the systems since the intervention in 1985. To provide a coherent framework for collecting and recording the interview information, the NIIS team designed a set of questions that focused on the governance and operation of the systems. Colleagues who conducted the interviews were required to write up each interview following the checklist format, which allowed us to code the interview information systematically for QCA analysis.

3. The values of the four-point scale are (1) abundance, (2) limited, (3) scarce, (4) non-existence.

4. In terms of the average tail-end intensity, there was a slight increase from 244.22 percent to 246.16 percent during the period, but the difference is not statistically significant. The average cropping intensity dropped from 246.16 percent in Time Slice 2 to 241.76 percent in Time Slice 3; the drop is statistically significant at the 0.1 level. It suggests that, generally speaking, the intervention effect on agricultural productivity could not be sustained.

5. While theories could help identify broad mechanisms of how contextual variables affect human choices, theories usually cannot predict the direction of these mechanisms, as well as how different mechanisms interact with one another to result in particular outcomes.

6.   The exchange rate in 1998 was US$ 1 = NRs 67.6.
7.   The Time Slice 2 data shows a similar pattern as the data for Time Slice 1. The ethnic composition for Time Slice 3 has missing values that prevent us from drawing conclusive patterns. Despite that, the data of Time Slices 1 and 2 are sufficient to support the argument that ethnic issues were not a confounding factor affecting irrigation performance.
8.   The dichotomized nature of the causal conditions is not a function of the QCA method but of the researcher. QCA can also handle continuous variables.
9.   For Equation (4.2), both the solution coverage and the solution consistency are 1.
10. For Equation (4.4), the solution coverage is 0.86 and the solution consistency is 1.

5 Post-intervention dynamics in 2008: focusing on two success and two failure cases

INTRODUCTION

In the previous chapter, we analyzed the configuration of success factors that influenced the 19 irrigation systems in the Indrawati River basin affected by the innovative WECS/IIMI action-research project that were recorded in 1999 and verified in 2001. Between 2001 and 2006, a range of difficulties and conflicts surfaced in all of Nepal due to the Maoist insurgency.

The study area was no exception. Farmers had to cope with these difficulties, and several mediation mechanisms were devised by community members in order to cope with the adverse effects of conflict. First and foremost, the farmers had to protect local resources by effective use of their own social capital networks. They used their embedded reciprocity through informal mechanisms in order to keep water resources away from the political root cause of conflict. Thus, due to the Maoist insurgence, locals had to live in extremely difficult situations that in turn impacted the farmers' livelihood and irrigation management. The farmers developed their own coping mechanism to deal with the parallel existence of the Maoist 'regime' and government authority at the village level. The farmers followed the government army, who would patrol in the villages during the daytime. On the other hand, during the nighttime Maoist rule, the farmers would avoid the army patrolling the main road by using the safety of the irrigation channels. The embeddedness of the social network helped to deal with the dual regime difficulties with minimum loss of human life.

The second challenge the farmers faced during the conflict period was the scarcity of able-bodied labor required for the regular and emergency repair and maintenance of their systems. The younger population of the community had either joined the Maoist revolutionary force or the government army in order to protect themselves, or had fled to the capital city of Kathmandu or to India in search of an alternative livelihood. As a result, only the elderly and female members were left behind, and the irrigation systems suffered repair and maintenance problems. New modes of cooperation rules had to be devised.

The third problem farmers faced was the unavailability of the small amount of repair and maintenance assistance provided annually by the Irrigation Development Division/Irrigation Development Subdivision (IDD/IDSD) and the local development fund provided to the Village Development Committee (VDC) by the District Development Committee (DDC) through the Ministry of Local Development. These funds were diverted by the District Administration for enhanced security of the district headquarters and local infrastructures rather than for repair and maintenance.

In this chapter, we draw on a return visit in 2008 to all of the 19 systems assisted by the WECS/IIMI intervention.[1] We will address two related issues. First, we examine overall changes in various aspects of irrigation with a special focus on changes in institutional arrangements related to operation and maintenance of these systems. We focus on how these systems have survived the conflicts and how locals are managing systems after the restoration of peace. In brief, we discuss overall changes that have occurred in regard to additional external assistance after 2001, if any, and institutional arrangements for irrigation management, including rule changes in the systems during this period. In the second half of the chapter, we select two of the most successful and two of the less successful systems for detailed investigation of the physical condition of the systems, of Water User Associations (WUAs), of agricultural performance, and the impact of the recent conflict in Nepal as governing factors for rendering them more successful and less successful.

OVERALL CHANGES IN VARIOUS ASPECTS OF IRRIGATION MANAGEMENT

As discussed in Chapter 4, after the WECS/IIMI intervention, some of the 19 systems received additional assistance mainly from two sources: VDC funds and the annual repair and maintenance support from IDD/IDSD.[2] These systems received assistance mainly for repair and maintenance of the canal. Similarly, some support was allocated to improve the structure of headworks and canal linings. Most of these efforts were provided prior to 2000 before the Maoist conflict. Between 2000 and 2008, little effort was made to assist in the improvement or maintenance of these systems, including the regular annual support provided by the VDC and IDD/IDSD. Due to these events, even those systems that needed repair and emergency maintenance had to find their own sources of funding from among the users or the condition of their systems would worsen.

After the WECS/IIMI intervention, several changes occurred in different aspects of institutional arrangements for irrigation management. The leadership, selection process, and changes in leadership patterns after intervention varied in different systems. During our follow-up visit in 2008, we found that 13 of the 19 systems had regular meetings with WUA functionaries and attended annual general meetings. The remaining six systems, however, rarely met, especially during the conflict period. Nearly half of the systems registered their irrigation systems with the District Administrative Office (DAO), which is a requisite for initiating any multifunctional cooperative initiative. But these processes could not continue during the conflict period.

Farmers in some of the systems have formulated different rules regarding system operation, and repair and maintenance, but in some cases no rules exist. That is, some rules are in written form and others, only through verbal understanding. Overall, as mentioned in Chapter 4, the WECS/IIMI intervention triggered rule formation in many systems, but the rules have not been implemented and followed effectively in all the systems and also vary across the systems.

During the conflict period, due to the lack of an effective enforcement mechanism as well as farmers' concern for the safety of their family, an inherent unanimity existed among the irrigators to make the latter their top priority. Other matters, such as rule-following, would be dealt with as an internal matter among the members. Following the popular uprising in April 2006, the country ended an 11-year conflict between the government and the Communist Party of Nepal (Maoist) with the signing of the Comprehensive Peace Agreement between the Seven Party Alliance (SPA) and the Communist Party of Nepal (Maoist). Though the conflict had devastating effects across the country with more than 13 000 people killed, and significant damage to critical infrastructure, reduced mobility and access to markets, which affected livelihoods and caused the loss of government services for much of the country, none of the WECS/IIMI-supported irrigation systems were reported to be adversely affected in physical terms or in a perceivable manner as a result of the conflict. The water users, who had been actively involved in the operation and maintenance of their irrigation systems, stated that it was mainly due to their solidarity regarding maintenance that the conflict did not have adverse effects on their physical systems.

During our visit to the systems in 2008, we inquired about system maintenance activities with a special focus on the involvement of users in maintenance and monitoring mechanisms to increase rule-following regarding maintenance. Overall, we found that, in all but a few cases, the users were involved in the operation and maintenance of their systems. In regard to operation and maintenance, the major-ity of the systems kept records, and occasionally imposed fines on those who did not participate in the operation and maintenance. In seven out of 19 systems, however, they did not maintain attendance records, nor impose fines on defaulters. The operation and main-tenance mechanism also varied across the systems. Farmers had relaxed participation in operation and maintenance and in record-keeping, choosing not to add an additional source of conflict that would further divide the community.

In half of the irrigation systems with an abundant quantity of water and high irrigation performance, no major conflicts were reported. Some conflicts related to water management did occur in the other half of these systems. The nature of the problem, those involved in the conflict, and the way the conflicts were managed varied across the systems. Generally, the conflict was among the farmers, but in several systems the most prominent conflict was between farmers and mill owners. In some cases, conflict erupted among irrigation systems due to sharing the same source of water. Another type of conflict present in the area was between farmers at the head and tail end of their systems. In a few cases of conflict between systems, external agency intervention was required. During the conflict, however, conflict-resolution mechanisms were restricted to informal social networks, be it within or among the irrigation systems. By reporting either to government official channels or the Maoist leaders, the farmers suffered more from the verdicts. Thus, they tended to take potential conflicts into their own hands as infor-mal social pressures were still respected by the farmers, and there was less threat to the continued survival of their systems.

During the time of conflict, the majority of the issues relating to water disputes were settled through dialogue among the conflicting parties, thus avoiding both Maoist and government forces. Similar strategies were reported in community-managed forest management in other Middle Hills districts of Nepal (Karna et al., 2010).

MORE SUCCESSFUL AND LESS SUCCESSFUL SYSTEMS DYNAMICS

Providing assistance with the intention of enhancing irrigation performance obviously seeks the sustainability of both irrigation systems and the social systems that have a direct bearing on agricultural performance (Ostrom, 2005; Vermillion, 2005). The second part of this chapter, therefore, seeks to assess the dynamics of rehabilitated farmer-managed irrigation systems (FMIS) and the factors responsible for causing positive and negative changes over time. As discussed earlier, an intervention, be it externally or internally initiated, is a necessary condition to enhance irrigation performance. One-shot initiatives alone are not sufficient to generate long-term sustainability of irrigation systems. Further, some systems that might be robust and sustainable during normal periods of social and political stability will face different problems during a period of conflict and instability and may then be exposed to social and ecological disturbances.

Based on the evaluation of condition and performance of the irrigation systems during our field visit in 2008, we discuss two irrigation systems in each of the two performance categories of irrigation systems so as to present the dynamics of the post-intervention situations in general, and coping strategies during conflict in particular. As mentioned in Chapter 4, we rated two irrigation systems as more successful (Dhap Kulo [system 4] and Majh Kulo [system 15]) and two as less successful or failure cases (Magar Kulo [system 8] and Dovaneswar [system 7]). Magar Kulo was identified in our over-time analysis in Chapter 4 as improved but fluctuating in terms of tail-end water availability and overall water adequacy. This system had improved greatly from Time Slice 1 to Time Slice 2, but began to deteriorate from Time Slice 2 to Time Slice 3. Due to the absence of any strong leadership since 1998, and more recent conflict over the diverse political ideologies among the users of the system, it has not been able to perform effectively and has become a failed system. Our qualitative descriptive analysis focuses primarily on four major aspects: physical condition of the irrigation systems, WUAs, agricultural performance, and the impact of the recent conflict in Nepal as major factors affecting their performance.

The Successful Systems

Dhap Kulo

This system was constructed in 1897 on a contract basis by the local people of a nearby village called Bans Khark. In the beginning, the system was used to irrigate only one field, which belonged to a retired army officer. The name of the irrigation system was Subedar (Army Major) Kulo. Other users were denied access to water. After nearly ten years of individual irrigation use, the name of the irrigation system was changed to Dhap Kulo, and other users were allowed to use water through regular participation in annual repair and maintenance. The system has been registered since 2002 with the Water Resource Committee and District Cooperative Office (Chautara, Sindhupalchok). In 1988, the designed area was 50 ha, which was expanded to 60 ha by 2006. During 2008, there were approximately 350 households receiving irrigation water from this system.

The system has been operating continuously since the WECS/IIMI assistance for monsoon and spring rice as well as for wheat, mustard, vegetable crops, and potato during the winter. In 1999, the Rural Energy Development Project (REDP) of the UNDP provided NRs 100000 for repair and maintenance of the canal linings. In 2002, the WUA was converted into a cooperative (registered with the Department of Cooperatives in March 2002 as 'Handi-Khola First Micro Hydro-Electricity Cooperative Limited'). In 2006, a large tract of upland area previously not irrigated was converted into irrigated land, thereby increasing the command area from 60 ha to 75 ha with the mobilization of the users' own resources. The cropping intensity on the land under this irrigation system is 300 percent, as there is an abundance of water available at the source as well as in the canal throughout the year.

Majh Kulo

Majh Kulo was first constructed in 1965 with the initiation of the farmers of Wards 4 and 5 of Shikharpur VDC. During the time of our field visit in 2008, the same executive committee that was formed in 1989 was providing leadership and was also responsible for the operation and maintenance of the scheme. The committee is composed of nine members. After the WECS/IIMI intervention, the Second Irrigation Sector Project (SISP), funded by the Asian Development Bank (ADB) in 1996–97, provided NRs 800 000 for masonry lining of about 1 km, repair of leakage points, and construction of an aqueduct near the source. The labor contribution worth NRs 400 000 was borne by the water users themselves. With these improvements, the source is now in good shape and little need exists for its repair and maintenance. In 1998, the DDC provided 60 gabion boxes, the VDC provided NRs 17 000 for filling the gabion boxes, and farmers contributed labor worth NRs 17 000. This new work was basically for repairs and linings of the canal as well as blocking any leakages in the canal, rather than for increasing the irrigated area. The WUA also maintained records of meetings and attendance of its members who contributed labor and cash for construction of new additional canal networks. The scheme has been registered with the DAO in Chautara – the headquarters of Sindhupalchok district in 1992.

The scheme initially provided irrigation water to 123 households, which increased to 150 after the assistance provided by SISP in 1996–97 and DDC in 1998. In 2008, 175 households irrigated 60 ha of land in Wards 4 and 5 and some parts of Ward 6 with nearly 0.33 ha of land per household. The scheme is operating well for monsoon rice, spring crops like maize, and winter crops such as mustard, wheat, potato, and other seasonal vegetable crops (cauliflower, radish, cabbage, leafy vegetables). The system has been operating continuously for three seasons (summer, fall, and winter). Though the quantity of water tends to decline during winter, it is still adequate for winter crops.

Before the conflict period, WUA meetings were held twice a year on a regular basis (before rice planting in June and before sowing winter crops in November). In addition, meetings were also held as needed with minutes of the meetings maintained. The canal is cleaned twice a year by the users, just before the monsoon rice season (June) and before the winter crop season (November). Almost all the users participated in the canal-cleaning activity. Most farmers (members of WUA) cooperated and participated in the maintenance of the canal. The WUA employed a *Paani Paale* (water monitor) to look after the system on a regular basis. During the conflict time, and even at the present time, only one-fourth of the total water users remained active, and they have not been able to support the *Paani Paale* even after the restoration of peace. Meetings of the executive (main) committee are held before they invite the farmers who are general members to discuss the operation and maintenance issues during the general assembly. When they have changes to propose to the existing policies and rules, they are discussed in the general assembly. A majority needs to approve the new rules and regulations related to the irrigation system and WUA management issues.

Discussion of Successful Irrigation System Dynamics

Recognized leaders exist in both of these successful schemes. The leaders are selected by consensus among the users (of water or of electricity). Meetings are held every month in connection with the micro-hydroelectricity scheme. During the meetings, the operation and maintenance of the irrigation system is discussed. In the case of Dhap Kulo, members of the micro-hydroelectricity scheme and representatives of VDC and REDP participate in the meetings. They now have rules made with the commencement of the micro-hydroelectricity in a cooperative pattern for distribution of electricity and operation and maintenance of the scheme, and the rules are well implemented. Besides the increment in agricultural production and productivity, the generation of electricity and operation of power-operated mills are additional achievements as a result of the effective and efficient management of the watercourses.

Condition of the systems during the conflict
In Dhap Kulo, neither the government force nor the insurgents made any attempt to disturb the functioning of the scheme. Majh Kulo of Shikharpur experienced some mild disturbances during this period. Movement of the users, even for the purposes of canal operation and maintenance, was not feasible because of frequent patrolling by the warring parties along the main canal, which in fact, was used as the route for their own purposes. The farmers worked to operate and maintain the canal only during the daytime (from 6:00 a.m. to 6:00 p.m.). At night, irrigation activities along these canals were too risky. Despite all this, the system continued to operate, but there was no systematic distribution of water as before. The irrigation water

appropriation activity was more on an individual basis rather than following the rules of the WUA.

During the conflict period, the users did not contribute labor for maintenance due to both a shortage of labor and because they feared for their lives. Very few irrigation service fees were paid. Due to lack of funds, the committee could not afford to pay fees to the watchmen, and since that time, the committee has been unable to employ any watchmen. It was not possible to conduct meetings and organize gatherings due to threats from the warring parties and frequently imposed curfews. The water users were scared; insurgents barred them from paying irrigation service fees, water monitors were not allowed to allocate and distribute water in the field, and the WUAs could not enforce the rules. Unlike in the past, the water users could not be mobilized for operation and maintenance of the canal due to a politically unfavorable situation. The leaders were only able to continue operating the system without using formal mechanisms and relying on the informal social network, and their non-political services in the past were honored by both insurgents and users alike. Therefore, despite the distressing aspects of this period, the water flow in the canal did not stop and the farmers continued their farming activities. Overall, there was no major effect on the operation and sustenance of the WUA nor was there any effect on the water supply/distribution system (at the head, middle, and tail of the system).

In Majh Kulo, only a fraction of the committed water users continued their routine work of operation and maintenance of the scheme in spite of the threats that came from the insurgents' side due to the differences in political ideology. Therefore, no modifications and changes in the governance mechanism (WUA) were felt to be necessary during the conflict period. In the case of Dhap Kulo, the committee implemented the cooperative's rules and regulations as the WUA had been converted into a cooperative institution.

There was a relative shortage of labor during the conflict because some people in the village fled to other places due to threats from the warring parties. A majority of elders and others who had no alternative stayed back in the village. The majority of those who stayed behind were involved in the operation and maintenance of the canal despite the threats and fear for their lives. As mentioned earlier, however, there was no overall reduction in the operation and maintenance activities.

Situation after peace agreement

The successful irrigation systems have been fairly stable. Their canal linings have not suffered major leakage and seepage problems. This is mainly due to the water users immediately fixing the canal as well as the headworks whenever any problems appeared. The physical system is performing satisfactorily, even in terms of water allocation and distribution. In terms of cropping intensity, it had attained 300 percent intensity in both systems. In recent years, there has been a diversification of crops. Farmers have started growing more vegetables, potatoes, and mustard, but have cultivated less wheat because of the occurrence of diseases.

The WUAs are functioning without any disturbances. The cooperative organization of Dhap Kulo has been performing much better with the production of electricity for the local people and generating a substantial amount of income from the sale of the power, besides operating the canal. Though the WUA of Majh Kulo could not be as active as it had been earlier during the conflict period in terms of rule enforcement and rule compliance, it has been gaining momentum gradually over time due to improvement in the peace and security situations in the country.

Therefore, there seems to be future potential for further improvement of system performance for Dhap Kulo and Majh Kulo. This is because of the availability of water and the capability of the Water User Committee (WUC) to harness available resources, including water. Recently, the VDCs located in Dhap Kulo and Majh Kulo have been connected with the road network. This provides the farmers and agro-entrepreneurs with market access. This development is likely to create a favorable environment for the farmers to diversify and commercialize their farming activities.

For the economic sustainability of their scheme, the WUC of Dhap Kulo plans to generate and supply power for daytime use (at present, the power supply is only for nighttime). Furthermore, the committee plans to generate electricity at a higher volume and sell the surplus to the non-members of the cooperative. As landslides interrupt the water flow in the canal and power generation, the committee intends to plant trees in the slide-prone area under the environment development program. The members of WUAs are looking ahead to further strengthen their institutions and the physical system. For this they intend to request help from the government, non-government organizations, and donor organizations.

Key factors of success

Based on a qualitative evaluation of these two most successful irrigation systems, there are important factors that are common to these systems. While the dependable sources of water for irrigation have been a major factor, an equally important factor is the multi-functional uses of water, thereby increasing water productivity and gaining additional economic benefits. Another important factor has been the institutional robustness and the willingness of users to find ways of cooperating during the Maoist insurgency period. The strong solidarity among the water users that fended off political interference even during the conflict time has certainly added to their success. This has been possible due to the existence of a strong institutional framework for managing the system. An equally important factor is the presence of a strong physical structure. Majh Kulo also continued to receive support from an external source (Rural Energy Development Program and District Agriculture Development Office's Small Irrigation Program) including credit support (investment) for various enterprise programs to the users.

The Less Successful Systems

Dovaneswar Kulo

Dovaneswar Kulo, situated in Thangpaldhap VDC and constructed in 1979, served ten households of Wards 3 and 4. The system was initiated by the local farmers who realized the need for irrigation water for farming activities. The initial design area was for 12 ha, whereas the actual irrigated area has been only 6 ha, thus indicating the reduced performance of the scheme. As discussed in Chapter 4, although in earlier years there was a cropping intensity of 300 percent and no change in the technical efficiency of the system had occurred, during and after conflict this irrigation system did not provide water for spring and winter season crops. Moreover, the irrigation system does not cover the whole area during these seasons. After the WECS/IIMI intervention, an area in the head portion expanded but some tail-end farmers did not receive water. This system never received assistance from any external sources besides the WECS/IIMI intervention.

After the intervention, no changes have occurred in the institutional arrangements for irrigation management. The same committee that existed at the time of WECS officially exists but is no longer working. No meetings were held after the project was completed. Unlike many of the other systems involved, this scheme was never formally registered with any governmental authorities. The leadership has not been effective, and no rules exist for irrigation management and operation. The system has been operated and managed as a result of verbal understandings and no system exists to impose fines on those who do not participate in the operation and maintenance.

The farmers themselves maintain and operate the system for a rice crop during the monsoon season, but no records exist of those involved in maintenance. There is no improvement in agricultural productivity; rather, it has decreased over time. The condition of the main canal is getting worse due to landslides caused by heavy rainfall. No leadership exists to take initiatives for the management of the system and the WUA.

Magar Kulo

Local farmers constructed the system in 1895. For almost a century, the system had been functioning in a disunited manner. Farmers who were in need of water came together in small groups of three to five members and diverted the water to their fields. There was no WUA until the WECS/IIMI rehabilitation. Since the demise of the chairman of the Construction Committee in 1998 during the WECS/IIMI rehabilitation, no WUC has existed. The farmers kept records only during the WECS/IIMI rehabilitation time. Therefore, the irrigation system has long been in operation on an individual basis, without a WUA or a leader. When farmers are in need of water, they go together in groups of three to five to bring back water for their farms. There is complete absence of any organized activities for the operation and maintenance of the system. In 2000, the VDC provided NRs 10 000 for the repair of the canal damaged by the flood. Thereafter, this scheme did not receive any support or assistance from external sources.

Due to the lack of proper care and maintenance, the tracks and linings of some portions of the canal have been disappearing over time. In addition, the agricultural productivity of the system has also decreased over time. Only two crops (summer rice and wheat or potato in winter) are grown with less than 100 percent intensity, which is down from the 140 percent level of cropping intensity achieved between 1991 and 1999 immediately after the WECS/IIMI intervention, as presented in Chapter 4. The tailenders have almost stopped receiving water, leading to reduced intensity, production, and productivity.

At present, the system covers a total irrigable area of 140 ha. The land under irrigation was expanded to 143 ha with the WECS/IIMI intervention. After the completion of the rehabilitation work, the irrigation system had operated at a satisfactory level with a continuous flow of water throughout its distribution system during part of the year. The system did not operate continuously and smoothly throughout the year even after the rehabilitation work was completed. The system operated only for monsoon rice, which was primarily due to the availability of rainwater. Rice is the primary crop grown during the monsoon season. For growing monsoon rice, farmers in need of water usually go in groups of four to five to clean the canal and bring back water for their farms. The same is true for wheat and mustard in winter, if any farmers planted their fields.

The farmers reported that since 1997, the system has not been in proper operation. Numerous (about 35) minor streams cross the main canal and the distribution canal (from the source to the fields). These streams often wash away the canal linings, especially during the monsoon. Thus, the system is operating on an individual basis without a WUA for operation and maintenance, without a sense of ownership of the system, no rules or regulations to control the water users, and water is being appropriated by small groups that can organize to divert water to their land. Landslides have also caused damage to the canal near the source. The highly fragile physiographic condition around the irrigation system is also one of the main reasons for reduced efficiency of the system. The system is often affected by landslides, especially the headwork and distribution canal located along the hill slopes.

Despite adequate water at the source, the approximately 5 ha area that lies around the tail end of the canal does not receive water at all because of the poor physical condition of the canal. This means the land under irrigation was reduced following the WECS/IIMI intervention. While there has been an addition of 1 ha of land around the head part of the canal, it does not receive irrigation water. The farmers reported that although about 30 percent of the total water in the source is delivered toward the tail end of the canal, due to conveyance losses caused by leakages, the tailenders still do not receive a sufficient amount of water for paddy cultivation.

Conditions of the systems during the conflict

Disturbances were experienced in the functioning of the WUAs. The 11-year-long conflict between the government and rebel forces brought the functioning of the WUA of Magar Kulo of Bhotenamlang VDC to a standstill and is a reflection of the impact of the power struggle between the two forces. It was not possible for the WUA to continue after the death of its chairman. Therefore, a vacuum was created due to lack of effective leadership acceptable to both opposing political views to enforce the rules and regulations. As a result, during the conflict period there was a situation of anarchy. Since many people fled or out-migrated due to threats from the warring parties, this situation not only created a void in leadership but also resulted in a serious lack of human resources to take care of the WUA and the physical system.

Situation after peace agreement

No WUA has existed since the demise of the then chairman (of the construction committee during the WECS/IIMI rehabilitation) in 1998. Thus, the irrigation system has been in operation on an individual basis due to the absence of a WUA. Obviously, no activities for operation and maintenance of the system are carried out in the form of collective action. Another complex aspect for this irrigation system is the large group size, that is, more than 400 household members. It has always been difficult for them to come together for any activity related to the irrigation system, be it physical or social. During 1991–98, the committee used to call for canal operation and maintenance (in the immediate period following the WECS/IIMI intervention), but that is not feasible now. The canal does not operate on a regular basis. Moreover, the canal linings are disappearing over time.

Since there is no WUA, no meetings are held regarding the irrigation system and/or reactivating the WUA. No rules were ever put in writing for the operation and maintenance of the system. From the beginning, they had only verbal agreements or understanding about the operation and maintenance of the system. Before the conflict in Nepal, the committee used to call for canal operation and maintenance (the immediate period following the WECS/IIMI intervention).

Conflicts/disputes occur mainly at the time of growing rice seedlings and rice transplanting during the monsoon season. Conflicts with the water mill owner also take place from March through June, as the mill owner diverts water from the main canal by breaking the structure (lining). Any conflicts are usually resolved through dialogue and understanding between the disputants.

The farmers still follow the traditional farming system with rice and wheat as major crops, rarely using high-yield farming technologies. Production/productivity has been reported to be decreasing over time, owing to the inadequacy of water. Only two crops (summer rice and mostly wheat in winter) are grown in the command area with less than 200 percent intensity. The decrease in cropping intensity could also be attributed to lack of management of the irrigation system for a third crop, as well as the occurrence of diseases. The farmers have now come to realize that their household income has been adversely affected due to the worsening of the irrigation system and the WUA.

There is substantial potential for increasing the irrigation and agricultural performance of the system since there is adequate water at the source and farmers are willing to cooperate and contribute. The farmers assert that a third crop, probably vegetables as a cash crop, could be introduced in the command area and income be increased considerably if the canal were well maintained and operated by reviving the WUC/WUA with strong leadership.

Factors contributing to lower levels of success

We visited the two less successful systems for detailed case studies of a very small size and a relatively large-sized irrigation system. Both of these systems have several common problems. These include the lack of effective leaders, polarization in political factions, weakness in both physical structures and agricultural innovations, and moreover, a complete absence of institutional arrangements for governance of these systems. These have happened due to lack of unity among the users to protect the system, political interference including the warring parties (eliminating the chance to organize and conduct meetings), lack of effective leadership, weak physical structure due to lack of maintenance leading to leakage, seepage, and linings damaged by landslides and overflow. In both systems, only the head and middle users receive adequate water. Further problems include weak on-farm management of water and distribution systems, lack of support from external sources after WECS/IIMI intervention, low productivity and intensity (due to higher altitude) as well as a lack of adopting other new agricultural technologies, displacement of local people during the conflict time, and choosing other occupations rather than returning to farming.

LESSONS LEARNED

From the successful cases, we re-emphasize the need for software aspects of irrigation governance mechanisms both for institutional and financial sustainability. Multiple uses of a resource make a system more sustainable, backed up by self-governance structures that in turn enhance unity and solidarity. To achieve long-term benefits, strong leadership is needed. In addition, facilitation of participatory/collective decision-making of various kinds by agencies at a local level through accommodation of diverse needs and interests of users will also tend to improve the outcomes. Through external credit investments, rural enterprises are making good profits. For this and for maintaining solidarity among the users, there is a need for some local people to be courageous and active. Harmonization/ integration with political groups at the local level (political harmonization) to avoid conflicts over leadership of the WUA is another important factor for long-term success.

Based on evidence from the less successful cases, our findings persistently indicate that the capacity development programs are essential in terms of development of institutional arrangements for rule formation and monitoring, support for leadership development and ability to cope with dynamic situations, and efforts to achieve multifunctional assistance including improved agricultural performances. These factors are more important in the design of assistance programs for the less successful systems. In order to make the less successful systems perform better, the important lessons that organizations wanting to help FMIS should emphasize relate to the sustenance of WUAs and institutional development programs, including improved agricultural programs to increase productivity. Leadership skills need to be supported through effective initiatives such as farmer-to-farmer training, which the WECS/IIMI program was able to do at an early period. Participatory decision-making involving various stakeholders is an important factor for the smooth operation of the systems. Equally important for inclusive development is harmonization with other systems' farmers and internal groups including different ethnicities, opposing political parties, women's groups, and various types of farmers (of small, medium, and large farms, as well as head- and tail-end farms).

NOTES

1. During the return visit to these systems, a team comprising Dr. Neeraj Joshi and Mr. Matrika Bhattarai spent two weeks in the study area in early 2008. In December 2008, Neeraj Joshi, Naresh Pradhan, Prachanda Pradhan, and Ganesh Shivakoti also visited selected systems for the case studies in this chapter.
2. After 1998, District Irrigation Offices were changed into IDDs/IDSDs.

6 Synthesis and conclusion

THE CHALLENGE OF OVERCOMING BEST PRACTICES

As discussed in Chapter 1, many international donors have searched over the past three decades for the 'best practices' related to improving irrigation performance. Two practices stand out as overly influential on the design of irrigation interventions around the world. Hiring external water engineers to design and construct up-to-date engineering infrastructure to replace the primitive structures that farmers have already constructed would be one of the two best practices. The second is developing an institutional template for how governmental agencies and farmers should be organized. Even though both of these templates have repeatedly been challenged (Yudelman, 1985; Chambers, 1988; Lam, 1996b), they still dominate international assistance to irrigation systems.

The WECS/IIMI intervention described in Chapter 3 tried hard to overcome these supposed panaceas. While engineers were charged with the responsibility of developing the specific blueprints for the remodeling of the systems in the Indrawati River basin, the farmers were initially charged with developing a list of the most important improvements. Then, the farmers were charged with contributing their own labor and if they were able to reduce the costs of improving the highest priority change, it was possible to move down the list of projects so as to achieve far more than would normally have been accomplished. Further, when the farmers themselves began to work in applying the initial engineering blueprints to their own systems, they were able to suggest various improvements that did not raise costs but made the systems potentially far more effective.

Second, the WECS/IIMI project strongly urged the farmers to craft a set of rules, but not a uniform set based on an externally designed institution. Farmer-to-farmer training exposed the Indrawati River systems to the rules and practices of several other farmer-managed systems in the hills of Nepal that had developed rules that worked in this environment. The farmers were urged *not* to just copy the rules of another successful system, but to learn some general principles that could be applied to the rules of their systems. As discussed in Chapters 4 and 5, having rules that the farmers themselves agreed to and followed was indeed a major factor separating the long-term successes from the systems that were not sustainably successful over time.

Given the small amount of funds invested in the WECS/IIMI intervention, that 13 out of 19 systems have improved their crop intensity and water at the tail end is an impressive accomplishment. The engineering interventions alone cannot account for the sustainability of some of the irrigation systems. The active involvement of the farmers has itself been a major factor in the long term, as discussed in Chapter 5. Thus, a core recommendation that we can make as a result of this long-term study of a very innovative intervention is that building social capital among the farmers is even more important than building engineering capital on the physical system. Within a few years of *any* external intervention, farmers will face new challenges in regard to biophysical factors, such as floods or droughts and landslides, or social-economic factors including major leadership changes or even major civil unrest, that they will have to determine how to overcome. Growing active leaders from within their communities and helping farmers to increase their own social capital is a very important challenge that is difficult to accomplish with a single intervention.

Considering the over-time dynamics facing farmer-managed irrigation systems (FMIS) in the Asian context, several new challenges have surfaced for the effective governance and management of irrigation systems in Asia. Research on irrigation management has accumulated a rich body of knowledge on irrigation governance, institutions, and management. This new knowledge has provided the basis for major changes in irrigation policies in the last several decades, including management transfer programs, assistance to farmer-managed irrigation systems, and irrigation financing. In the beginning of the twenty-first century, however, additional broader issues have surfaced that include: how to respond to the competition for water resources among different sectors; what aspects of institutional reforms deal with governance and management of water resources; and how irrigation management can be made pro-poor, responding to livelihood sustenance requirements.

In an effort to address these challenges thoroughly so as to provide a firm foundation for confronting them effectively, the following five themes need to be addressed.

1. The processes of globalization, industrialization, and urbanization are all generating immense pressures for a *transition* from earlier political, economic, and social institutions to new arrangements in all sectors.
2. *Competition* for resources – particularly water – will increase throughout the world over time, leading to immense conflicts unless substantial innovations occur.
3. *Institutional reforms* are among the most important innovations that are needed to meet these challenges.
4. *Markets* will be a more important aspect of water management in the future than they have been in the past.
5. *Strategic policies* are needed that are conducive to governing and managing water resources effectively in the light of transition, competition, and institutional and market reforms (Shivakoti, 2006).

There is also a need to identify and document the changes in the context of contemporary irrigation management through an assessment of how these macro-changes affect the incentives, opportunities, and constraints of farmers at the local level. Finally, an understanding of how and why farmers in different settings have adjusted, or failed to adjust, their local irrigation in response to the changing context is pertinent and discussed in Chapter 2.

BROAD LESSONS FROM STUDIES OF DIFFERENT MODES OF IRRIGATION

In Chapter 2, we presented and analyzed a large data set on diversely organized irrigation systems collected during the 1990s by colleagues at Indiana University and at the Institute of Agriculture and Animal Science (IAAS) in Rampur. These included 112 FMIS that had received interventions that involved high, medium, or low levels of farmers' involvement; 69 FMIS that had not received any external assistance; and 48 agency-managed irrigation systems (AMIS) financed entirely by the Nepal government and initially constructed with funds provided by international aid agencies.

Our discussion of the performance of irrigation systems included three dimensions: (1) the physical condition of irrigation systems, (2) water delivery, and (3) agricultural productivity. All of these three dimensions are important and they cannot be aggregated into one scale. As we point out, an irrigation system does not perform well if its canals are maintained effectively but its water delivery is unsatisfactory, or if the physical system has effective water delivery, but farmers are not able or encouraged to use the water efficiently to increase agricultural productivity.

We found that about 83 percent of AMIS had permanent headworks, but less than one-tenth of these systems were excellent in terms of physical condition. AMIS were also less able than FMIS to get adequate and predictable water to the tail end of their systems even though most of them are fully lined with permanent headworks. The existence of permanent headworks often exacerbates the difference between headenders and tailenders, leading to low levels of system performance. When they do not get adequate water, the tailenders are not likely to engage in maintenance of a system, nor follow the rules concerning water distribution and allocation.

We found it important to note that a greater proportion of AMIS operated without farmers' participation were inefficient as compared with the systems built and managed by the farmers themselves, whether or not they received external aid. This reflects the lack of understanding by the farmers being served of the administrative system and rules of AMIS. The rules made by the farmers collectively in FMIS facilitate the development of shared norms and emphasize the importance of self-reliance and cooperation.

The data and analysis presented in Chapter 2 provide useful information about the potential of diversely organized, externally assisted irrigation schemes and the factors that affect the performance of irrigation systems. Since many intervention projects focus more on the physical infrastructure of irrigation systems, they ignore the social infrastructure. Physical infrastructure definitely plays a key role. The social and institutional aspects, however, also have an important role to play in the improvement of irrigation system performance. In the irrigation schemes implemented without consultation of the potential water users, it is very likely that the users tend not to contribute labor in the operation and maintenance tasks to be performed on a collective-action basis.

AN INNOVATIVE INTERVENTION

In Chapter 3, we described an innovative effort to improve FMIS undertaken by WECS/IIMI with support from the Ford Foundation. The location for this action-research project was the Upper Indrawati River basin in Sindhupalchok District of the Middle Hills of Nepal. This action-research project self-consciously developed a set of procedures that they hoped would improve the overall operation of 19 FMIS so as to also improve the efficiency of the systems, the social capital that farmers could develop, as well as helping to increase the long-run sustainability of at least some of the systems affected by the intervention.

After a review of all irrigation systems in the 200 sq km project area, a reconnaissance/inventory preparation study was undertaken to determine the resource base of each system within the project area. On the basis of the inventory, potential candidates for assistance were identified using a variety of indicators. Finally, 22 systems were identified as candidates for improvement on the basis of having the possibility of command area expansion and extra water resources available. This led to a further appraisal regarding the need for physical improvement. Finally, 19 out of 119 systems were selected for assistance.

The next step taken by the team was to initiate a series of dialogues between the farmers from the 19 systems and agency personnel. A serious effort was made to obtain accurate information about the families being served, how strongly they were organized and their irrigation management practices. Critical areas that needed physical improvement were also identified. An initial rapport was also achieved between the farmers and project staff.

During the rapid appraisals, farmer training for irrigation management in each system was identified as a priority for the successful implementation of the project. Although the field supervisors assisted farmer management by advising committee members in group decisions, keeping records, and mobilization of labor, the result was not satisfactory and a system of farmer-to-farmer training was initiated. That enabled the farmers to observe how farmers on similar systems to their own handled a variety of shared problems and how they had organized themselves to be effective. Further, the project team discovered that the farmers had found it difficult to understand blueprints and design sketches. Thus, the design engineers were asked to use three-dimensional scaled models for explaining how the proposed structures looked and operated as well as pointing to such structures used in the other systems the farmers had visited.

Using a very moderate financial investment (about $150 per hectare), the project was able to expand the irrigated area commanded by the canals by more than 50 percent. The active participation of farmers in the planning and development of the designs and in the construction of new facilities substantially increased the value of the initial grant over what would have been needed if contractors had been used. Thus, the WECS/IIMI project used a different perspective than the 'best practices' recommended at that time by development agencies: that intervention should be a process of transferring resources to farmers by government agencies. The WECS/IIMI perspective was that it was essential to try to enhance farmers' ability to manage their own systems. Through intervention, farmers can be enabled to maintain their system and to engage in self-governing activities concerning appropriation and maintenance.

EVALUATING AN INNOVATIVE DESIGN FOR IMPROVING FMIS MANAGEMENT

In Chapter 4, we examined how improved engineering works configure with other factors to affect irrigation performance. Policy analysts have tended to evaluate interventions thought of as simple additive processes rather than complex configural processes. We attempted to decipher the complex, over-time processes started by the WECS/IIMI intervention that had tried to avoid imposing a 'best-practices' cure-all on local farmers.

We used an augmented version of our existing Nepal Irrigation Institutions and Systems (NIIS) database described in Chapter 2 to code the initial data collected by the WECS/IIMI team in 1985 about the systems before the intervention. This became 'Time Slice 1' and serves as the benchmark to comparing various outcomes with the situations prior to the intervention. Members of the NIIS research team visited the 19 systems included in the project to conduct a second round of data collection a few years after the intervention. This constitutes the 'Time Slice 2' data in our tables. The NIIS team visited the 19 systems again in 1999 (eight years after the second visit) to assess a longer-term impact, and this information is the 'Time Slice 3' data used in Chapter 4. Looking at the relationship between these three time slices provides one view of the net effect of intervention. We paid particular attention to five key measures of performance: size of irrigated areas, technical efficiency of irrigation infrastructure, water adequacy, tail-end cropping intensity and levels of deprivation in a system.

With these time slices, we had only three snapshots of an unfolding process and confronted the methodological problems of using conventional statistical analysis, which assumes that the effects of individual variables are independent from one another. It is often the case that it is the combinatorial effects of several variables that affect outcomes and these are rarely linearly related. To synthesize the data, we applied Boolean algebra and the Qualitative Comparative Analysis (QCA) developed by Ragin (1987), which allows for a more systematic distinction, between necessary and sufficient causal conditions (Rihoux, 2008).

The detail of our analysis of the experience of the WECS/IIMI intervention is presented in Chapter 4. We show that improving infrastructure improves technical efficiency only in the short run. Given the challenging environment of the Middle Hills of Nepal, maintaining a high level of technical efficiency by investing in infrastructure improvement alone cannot sustain a high level of technical efficiency.

Our analysis suggests that, in most of the 19 systems involved, improved technical efficiency of irrigation infrastructure did bring about a persistent improvement in water adequacy. The infrastructure improvement works undertaken by farmers enabled them to see for themselves how their effort could make a difference (and how to develop working relationships with one another). When they were able to keep up a good working order, they also achieved a high level of water adequacy. The WECS/IIMI intervention did help the farmers improve such abilities and skills and avoided making them dependent on external aid. Through farmer-to-farmer training and other means of getting the farmers involved, and identifying local leaders and rules that would work for them, the project did avoid the 'best-practices trap' of relying on externally designed infrastructure improvement.

The 19 systems did take different paths for self-organization. Some have thrived while others have failed to sustain the physical improvements achieved early in the process. Once again, we show that there is not a single recipe for developing human artisanship. As long as farmers are willing to maintain a certain level of collective action and local entrepreneurs exist to provide leadership, we show that it is possible for farmers to build on the momentum introduced by an interactive intervention to attain high levels of performance over time.

FURTHER LESSONS FROM AN INNOVATIVE INTERVENTION

In Chapter 5, we discussed two related questions. The first relates to the general patterns of development that occurred in the project area between 2001and 2008. While the last visit to all 19 systems for collecting data to enter in the NIIS database was in 1999, and the QCA reported on in Chapter 4 used that data as the third data entry, we were fortunate that colleagues made a supplementary visit in 2001 to verify the basic information that had been entered two years earlier. In 2008, colleagues were able to return to the project area and provide an update on the level of external maintenance that had been provided over time. The second question discussed in Chapter 5 relates to how farmers coped with the conflict between Maoist and the government's armies during the period from 2001 to 2008.

We found that systems varied in regard to both the level of external support they received toward the end of the last century and in regard to their own capacity to maintain their systems and to keep their rules enforced. When local leaders were able to build on well-accepted rules and a sense of general reciprocity, the irrigation system was relatively well maintained even without much external help. However, many of the FMIS were not able to continue active engagement. The farmers in these systems reduced their levels of participation in the maintenance of their systems and their record-keeping about who continued to make investments in the systems. Conflict also emerged among the farmers on some of these systems. Thus, the FMIS that had received WECS/IIMI intervention varied in regard to their level of continued maintenance of their rules and of their systems.

Further, the conflict between the Maoist and the government's armed forces led to a substantial reduction of labor that could be devoted to the maintenance of these systems. Most young members of farming families left their homes to join either the Maoist revolutionary or the government forces or fled to Kathmandu or to India to seek out alternative ways of making a living. Most of the 19 irrigation systems suffered substantial repair and maintenance problems and thus also needed to develop new ways of cooperating in the shadow of the conflict. The conflict also led to substantial reduction in the small amount of repair and maintenance funds provided by the District Irrigation Office or by the Village Development Committee.

The second part of Chapter 5 was devoted to an assessment of the long-term performance of FMIS receiving WECS/IIMI intervention almost three decades earlier. Two irrigation systems that were more successful and two systems that were less successful were chosen for in-depth analysis. In general, the WECS/IIMI intervention was found to have been an essential foundation for the sustainability of some of the systems that received the initial intervention described in Chapter 3. Thus, the intervention was a necessary condition for enhancing irrigation performance over time. The initial one-shot initiative alone was not an adequate measure for the long-term sustainability of the irrigation systems. Equally important are the follow-up by government offices, the rules and monitoring systems created by the farmers, and the concerted efforts by stakeholders toward making the systems sustainable over the long run from the farmers' perspectives.

AN AGENDA FOR EFFECTIVE INTERVENTIONS WITH FMIS

Accomplishments of the past two decades in understanding and modifying governance and management of irrigation systems have been considerable. However, the twenty-first century brings new challenges. Even in the remote area of Indrawati watershed, like everywhere else in Asia, population growth and urbanization have placed industrial and environmental water needs in competition with water previously allocated for food production and the ever-increasing need to use water more efficiently for food production. While management responsibility is being transferred to local user groups, property rights and rights to water often remain unresolved issues. The transfer of irrigation system governance and management has often overlooked the need to modify information systems and access to information necessary for responsible governance. Thus, the following three integrating themes will be of considerable importance over the next few decades:

1. Responding to competition for resources (water scarcity):

   – transfer of water for multifunctions and irrigated land from agriculture to other uses (municipal) and its impact on agriculture;
   – watershed depletion, water-quality degradation and water reclamation;
   – property rights: water, land, infrastructure (including the planning, information, and administrative requirements of these);
   – linking downstream and catchment stakeholders in watersheds and effective management of water resources (including information and communication requirements).

2. Accountability and new partnerships:

   – transfer of authority for irrigation system management;
   – changing government roles to regulation, provision of support services and capacity-building based on the principle of co-management and 'polycentric' governance integrating farmers' institutions of irrigation management;
   – new accountability mechanisms: service agreements, management audits, asset management plans (including information requirements) in light of multi-use of water;
   – redesigning government subsidies in light of the public–private partnership with enhanced ownership stakes made by local users and matching investments, transparent and agreed allocation criteria, incremental infrastructure improvement.

3. Reform, synergy, and economic productivity:

   – scheme-level Water User Association (WUA) federations and new opportunities for WUAs hiring their own agricultural/agribusiness development agents;

- new information/communication systems for market identification and networking;
- using new demand-oriented irrigation services to promote crop diversification and commercialization;
- expanding the economic niche of farmers beyond cultivation to agribusiness (input production and supply, crop processing, production of manufactured agricultural/horticultural products);
- joint monitoring for irrigation operation and multiple uses of water: diverse methods of data collection, storage, and processing into information, public access and sharing of information;
- transition of irrigation operation policy beyond water provision to cope with changing economic context and World Trade Organization requirements;
- exploring alternate mechanisms for governance and management of irrigation into the larger context of economic integration, competition for water and the need for water conservation and land resources;
- institutional reform of irrigation development agencies should imbibe the new values of socioeconomic changes of inclusiveness, gender concern, self-governing local water institutions, and poverty alleviation.

SOME CONCLUDING THOUGHTS

This has been an unusual volume stimulated by an unusual intervention project and our opportunity to observe patterns of organization and outcomes over time for 19 irrigation systems located in the Indrawati River basin in the Middle Hills of Nepal. The WECS/ IIMI intervention described in Chapter 3 was truly innovative. The amount of funds allocated to helping the farmers served by the 19 systems was very small when compared with most externally funded irrigation improvement grants. The amount of time and energy spent by the project team with the farmers, on the other hand, was substantial. The WECS/IIMI team used the small financial grant to obtain good engineering advice but did not stop after engineers submitted their drawings. The farmers on these systems were asked for serious advice as to what physical aspects of their systems needed the most attention. The farmers were also asked to provide significant amounts of labor so as to stretch the financial resources as far as possible while significantly upgrading the physical aspects of their irrigation systems, and to learn about the construction and maintenance of their own systems. Further, the farmers were asked to invest in improving their own institutional arrangements and participate in farmer-to-farmer training to learn how farmers, who did manage successful systems, had crafted their institutions.

Intervention experiences among the Asian countries teach us that assistance to farmer-managed systems should avoid making them more dependent on external assistance, and provide opportunities to the farmers to make decisions regarding the operation of the irrigation system based on their local knowledge and creativity. The success story of intervention in Asian countries tells us that intervention programs should be technically feasible, institutionally sustainable, and economically profitable. Instead of assuming the existence of 'the' most appropriate institutional arrangement that, when put in place, would automatically bring about specific desirable conditions, the innovative intervention focused on identifying imperatives for long-term sustainability, including effective feedback loops, use of local knowledge and problem-solving capacity, and creating a conducive setting for attaining these imperatives.

The WECS/IIMI intervention was a serious effort to devise an effective alternative to the then dominant 'best practices' of international assistance agencies to invest in improving the engineering works of irrigation systems. They understood that in addition to improving the physical works of an irrigation system, an equally important aspect of improved performance was improving the social capital of the farmers who would have to learn how to sustain any external investment or see the system erode over time. Implicitly, they recognized that irrigation systems existed within complex social-ecological systems (SESs) and that focusing only on the physical works of a system did not include the important social factors that also affect performance over time. Further, they understood that swamping the farmers with a substantial financial grant frequently led farmers to be dependent on external aid rather than develop their own skills for operating and managing a system over time.

Fortunately, colleagues associated with the Irrigation Management Systems Study Group (IMSSG) at the Institute of Agriculture and Animal Science (IAAS), Tribhuvan University,

Nepal, in collaboration with colleagues at the Workshop in Political Theory and Policy Analysis, Indiana University, USA, had already invested substantial time and energy devising, testing, and revising the Nepal Irrigation Institutions and Systems (NIIS) database. They had already collected data about irrigation systems across Nepal as reported in Chapter 2. They decided to follow the innovative WECS/ IIMI project by including data about the 19 systems before the intervention (1985) in the NIIS database. Then our Nepali colleagues returned in 1991 to gather and enter data about the systems for a second time slice, and for a third time slice in 1999. More qualitative data was collected in 2001 and 2008. Thus, an innovative scientific effort was made to follow over time those irrigation systems that had been involved in an innovative intervention effort.

It is no surprise that the 19 systems have taken different paths after the WECS/IIMI intervention, given their substantially different history and social-political backgrounds prior to the intervention. It is wonderful to record the capacity of some of the systems built upon the momentum of the intervention which have thrived over a long and difficult period. It is not unexpected to learn that other systems failed to sustain the physical improvements achieved early in the process. As we have stressed throughout this volume, it is naive to think that there is 'one best way' to intervene and improve irrigation system performance over time. Yet analysis over two decades does support a conclusion that farmers can build on an innovative intervention that provides training to them and not just physical improvements. But to achieve long-run improvements in performance, the farmers *themselves* need to engage in collective action over time and support local entrepreneurs who provide leadership and stimulate adjustments to change. Hopefully, this study will provide strong support for other innovative teams to design intervention projects well tailored to local circumstances and involving substantial training, drawing on knowledge acquired by successful farmer-managed systems in the region.

The experiences of the WECS/IIMI intervention teach us that irrigation systems are SESs. Therefore, an intervention has to have multidimensional features to address resources (water), physical infrastructures (canals and other control structures), as well as placing farmers in the driver's seat and creating appropriate governance procedures (irrigation institutions).

References

Acharya, B.N. (1990), 'Design issues in farmer-managed irrigation systems in Nepal: experiences in the hills of Nepal', in R. Yoder and J. Thurston (eds), *Design Issues in Farmer-Managed Irrigation Systems*, Colombo, Sri Lanka: IIMI, pp. 121–32.

Anderies, J., M. Janssen and E. Ostrom (2004), 'A framework to analyze the robustness of social-ecological systems from an institutional perspective', *Ecology and Society*, **9** (1), 18. Available online at: http://www.ecologyandsociety.org/vol9/iss1/art18/inline.html; accessed 11 April 2011.

Araral, E. (2005), 'Bureaucratic incentives, path dependence, and foreign aid: an empirical institutional analysis of irrigation in the Philippines', *Policy Sciences*, **38** (2–3), 131–57.

Araral, E. (2009), 'What explains collective action in the commons? Theory and evidence from the Philippines', *World Development*, **37** (3), 687–97.

Ascher, W. and R. Healy (1990), *Natural Resource Policy Making in Developing Countries*, Durham, NC: Duke University Press.

Axelrod, R. and M.D. Cohen (2000), *Harnessing Complexity*, New York: Basic Books.

Bagadion, B. (1987), 'Government intervention in farmer-managed irrigation systems in the Philippines: how research contributed to improving the process', in *Public Intervention in Farmer-Managed Irrigation Systems*, Colombo, Sri Lanka: IIMI and WECS of the MOWR, GON, pp. 265–76.

Baker, M. (2005), *The Kuhls of Kangra: Community Managed Irrigation in the Western Himalaya*, Seattle: University of Washington Press.

Barker, R., E.W. Coward Jr., G. Levine and L.E. Small (1984), *Irrigation Development in Asia: Past Trends and Future Directions*, Ithaca, NY: Cornell University Press.

Bauer, C.J. (1997), 'Bringing water markets down to earth: the political economy of water rights in Chile, 1976–95', *World Development*, **25** (5), 639–56.

Berkes, F. (2002), 'Cross-scale institutional linkages: perspectives from the bottom up', in E. Ostrom, T. Dietz, N. Dolšak, P. Stern, S. Stonich and E. Weber (eds), *The Drama of the Commons*, National Research Council, Washington, DC: National Academies Press, pp. 293–325.

Berkes, F. and C. Folke (1998), *Linking Social and Ecological Systems: Management Practices and Social Mechanisms for Building Resilience*, Cambridge, UK: Cambridge University Press.

Bhattarai, L.N. (1990), 'Low-cost assistance to small farmer-managed irrigation systems', in R. Yoder and J. Thurston (eds), *Design Issues in Farmer-Managed Irrigation Systems*, Colombo, Sri Lanka: IIMI, pp. 63–76.

Bowles, S., S.N. Durlauf and K. Hoff (eds) (2006), *Poverty Traps*, Princeton, NJ: Russell Sage Foundation.

Bruns, B. and S.D. Atmanto (1992), 'How to turn over irrigation systems to farmers? Questions and decisions in Indonesia', *Irrigation Management Network Paper* No. 10. London: ODI.

Burns, R.E. (1993), 'Irrigated rice culture in monsoon Asia: the search for an effective water control technology', *World Development*, **21** (5), 771–89.

Carlson, J.M. and J. Doyle (2002), 'Complexity and robustness', *Proceedings of the National Academy of Sciences*, **99** (Suppl. 1), 2538–45.

Carruthers, I. (1981), 'Neglect of O&M in irrigation: the need for new sources and forms of support', *Water Supply and Management*, **5**, 53–65.

Cernea, M.M. (1987), 'Farmer organization and institution building for sustainable development', *Regional Development Dialogue*, **8** (2), 1–24, Nagoya, Japan: United Nations Centre for Regional Development.

Chambers, R. (1988), *Managing Canal Irrigation: Practical Analysis from South Asia*, Cambridge, UK: Cambridge University Press.

Coward, E.W., Jr. (ed.) (1980), *Irrigation and Agricultural Development in Asia*, Ithaca, NY: Cornell University Press.

Curtis, D. (1991), *Beyond Government: Organisations for Common Benefit*, London: Macmillan.

de los Reyes, R.P. (1980), *47 Communal Gravity Systems: Organization Profiles*, Quezon City, Philippines: Ateneo de Manila University, Institute of Philippine Culture.

de los Reyes, R.P. and S.M.G. Jopillo (1988), 'The impact of participation: an evaluation of the NIA's communal irrigation program', in F. Korten and R.Y. Siy Jr. (eds), *Transforming a Bureaucracy: The Experience of the Philippine National Irrigation Administration*, West Hartford, CT: Kumarian Press, pp. 90–116.

Dietz, T., E. Ostrom and P. Stern (2003), 'The struggle to govern the commons', *Science*, **302** (5652), 1907–12.

Dinar, A., M.W. Rosegrant and R. Meinzen-Dick (1997), 'Water allocation mechanisms: principles and examples', Washington, DC: The World Bank, Agriculture and Natural Resources Department.

Easter, K.W. (1986), *Irrigation Investment, Technology, and Management Strategies for Development*, Studies in Water Policy and Management No. 9, Boulder, CO: Westview Press.

Easterly, W. (2001), *The Elusive Quest for Growth: Economists' Adventures and Misadventures in the Tropics*, Cambridge, MA: MIT Press.

Easterly, W. (2006), *The White Man's Burden: Why the West's Efforts to Aid the Rest Have Done So Much Ill and So Little Good*, New York: Penguin Press.

Evans, P.B. (1996), 'Government action, social capital and development: reviewing the evidence on synergy', *World Development*, **24** (6), 1119–32.

Evans, P.B. (2004), 'Development as institutional change: the pitfalls of monocropping and the potentials of deliberation', *Studies in Comparative International Development*, **39** (4), 30–52.

FAO (Food and Agriculture Organization of the United Nations) (2003), *World Agriculture Towards 2015/2030: An FAO Perspective*, London: FAO/Earthscan.

Gautam, U., N.K. Agrawal and R. Subedi (1992), 'Nepal managing large surface irrigation projects: a participatory review', Study Document NEP/89/006, Kathmandu, Nepal: Department of Irrigation and Consolidated Management Services.

Gibson, C., K. Andersson, E. Ostrom and S. Shivakumar (2005), *The Samaritan's Dilemma: The Political Economy of Development Aid*, Oxford, UK: Oxford University Press.

Gill, M.A. (1991), 'Farm-level water management systems (public and private)', in Asian Productivity Organization (ed.), *Farm-Level Irrigation Water Management: Report of a Study Meeting in Lahore, Pakistan*, Tokyo: Asian Productivity Organization, pp. 79–87.

HMG/N, Department of Irrigation and ADB/Manila (1997), *Operational Procedural Manual*, vol. 1, Main Text, Second Irrigation Sector Project, Loan No. 1437 NEP, Kathmandu, Nepal: NIA Consult and East Consult.

HMG/N, National Planning Commission of Nepal (1994), *Irrigation Development in Retrospect: Search for a Breakthrough*, Kathmandu, Nepal: National Planning Commission.

Hussein, M., H.W. Khan, Z. Alam and T. Husain (1987), 'An evaluation of irrigation projects undertaken by AKRSP in the Gilgit district of northern Pakistan', in *Public Intervention in Farmer-Managed Irrigation Systems*, Colombo, Sri Lanka: IIMI and WECS of the MOWR, GON, pp. 237–64.

IIMI/WECS (International Irrigation Management Institute/ Water and Energy Commission Secretariat, Nepal) (1987), *Public Intervention in Farmer-Managed Irrigation Systems*, Colombo, Sri Lanka: IIMI and WECS of the MOWR, GON.

Janssen, M., J.M. Anderies and E. Ostrom (2007), 'Robustness of social-ecological systems to spatial and temporal variability', *Society and Natural Resources*, **20** (4), 307–22.

Johnson, S.H., III (1991), 'Status and progress of irrigation development: policy and programmes', in Asian Productivity Organization (ed.), *Farm-Level Irrigation Water Management: Report of a Study Meeting in Lahore, Pakistan*, Tokyo: Asian Productivity Organization.

Karki, J. (2001), 'Impact study on external assistance to farmer managed irrigation systems in Nepal', in U. Gautam and S. Rana (eds), *Challenges to Farmer-Managed Irrigation Systems*, Kathmandu, Nepal: Farmer Managed Irrigation Systems Promotion Trust, pp. 163–72.

Karna, B., G.P. Shivakoti and E.L. Webb (2010), 'Resilience of community forestry under conditions of armed conflict in Nepal', *Environmental Conservation*, **37** (2), 1–9.

Korten, F.F. (1987), 'Making research relevant to action: a social learning perspective', in *Public Intervention in Farmer-Managed Irrigation Systems*, Colombo, Sri Lanka: IIMI and WECS of the MOWR, GON, pp. 293–306.

Korten, F. and R. Siy Jr. (1987), *Transforming a Bureaucracy: The Experience of the Philippine National Irrigation Administration*, Quezon City, Philippines: Ateneo de Manila University Press.

Lam, W.F. (1994), 'Institutions, engineering infrastructure, and performance in the governance and management of irrigation systems: the case of Nepal', PhD dissertation, Indiana University.

Lam, W.F. (1996a), 'Institutional design of public agencies and coproduction: a study of irrigation associations in Taiwan', *World Development*, **24** (6), 1039–54.

Lam, W.F. (1996b), 'Improving the performance of small-scale irrigation systems: the effects of technological investments and governance structure on irrigation performance in Nepal', *World Development*, **24** (8), 1301–15.

Lam, W.F. (1998), *Governing Irrigation Systems in Nepal: Institutions, Infrastructure, and Collective Action*, Oakland, CA: ICS Press.

Lam, W.F. (2001), 'Coping with change: a study of local irrigation institutions in Taiwan', *World Development*, **29** (9), 1569–92.

Lam, W.F. (2005), 'Reforming Taiwan's irrigation associations: getting the nesting of institutions right', in G. Shivakoti, W.F. Lam, D. Vermillion, E. Ostrom, U. Pradhan and R. Yoder (eds), *Asian Irrigation Systems in Transition: Responding to the Challenges Ahead*, New Delhi: Sage Publications, pp. 346–64.

Lam, W.F. (2006a), 'Foundations of a robust social-ecological system: irrigation institutions in Taiwan', *Journal of Institutional Economics*, **2** (2), 1–24.

Lam, W.F. (2006b), 'Designing institutions for irrigation management: comparing irrigation agencies in Nepal and Taiwan', *Property Management*, **24** (2), 162–78.

Lam, W.F. (2011), 'Governing the commons', in M. Bevir (ed.), *The SAGE Handbook of Governance*, Thousand Oaks, CA: Sage Publications, pp. 501–17.

Lam, W.F. and E. Ostrom (2010) 'Analyzing the dynamic complexity of development interventions: lessons from an irrigation experiment in Nepal', *Policy Sciences*, **43** (1), 1–25.

Lam, W.F. and G. Shivakoti (2002), 'Farmer-to-farmer training as an alternative intervention strategy', in G. Shivakoti and E. Ostrom (eds), *Improving Irrigation Governance and Management in Nepal*, Oakland, CA: ICS Press, pp. 204–21.

Lam, W.F., M. Lee and E. Ostrom (1994), 'An institutional analysis approach: findings from the NIIS on irrigation performance', in J. Sowerwine, G. Shivakoti, U. Pradhan, A. Shukla and E. Ostrom (eds), *Proceedings of an International Workshop on Farmers' Fields to Data Fields and Back: A Synthesis of Participatory Information*

Improving irrigation in Asia

Systems for Irrigation and Other Resources*, Colombo, Sri Lanka: IIMI, and Rampur, Nepal: IAAS, pp. 69–93.

Lam, W.F., M. Lee and E. Ostrom (1997), 'The institutional analysis and development framework: application to irrigation policy in Nepal', in D.W. Brinkerhoff (ed.), *Policy Studies and Developing Nations: An Institutional and Implementation Focus*, Greenwich, CT: JAI Press, pp. 53–85.

Martin, E.D. and R. Yoder (1987), 'Conference overview', in *Public Intervention in Farmer-Managed Irrigation Systems*, Colombo, Sri Lanka: IIMI and WECS of the MOWR, GON, pp. iii–vii.

Medagama, J. (1987), 'State intervention in Sri Lanka's village irrigation rehabilitation program', in *Public Intervention in Farmer-Managed Irrigation Systems*, Colombo, Sri Lanka: IIMI and WECS of the MOWR, GON, pp. 215–32.

Miller, J.H. and S.E. Page (2007), *Complex Adaptive Systems*, Princeton, NJ: Princeton University Press.

Mitchell, M. (2009), *Complexity: A Guided Tour*, New York: Oxford University Press.

Molden, D. (2007), *Water for Food, Water for Life: A Comprehensive Assessment of Water Management in Agriculture*, London: Earthscan.

Moore, M. (1989), 'The fruits and fallacies of neoliberalism: the case of irrigation policy', *World Development*, **17** (11), 1733–50.

NPC (National Planning Commission) (1997), *Statistical Year Book of Nepal*, Kathmandu, Nepal: His Majesty's Government, Central Bureau of Statistics, National Planning Commission.

Ostrom, E. (1990), *Governing the Commons: The Evolution of Institutions for Collective Action*, New York: Cambridge University Press.

Ostrom, E. (1992), *Crafting Institutions for Self-Governing Irrigation Systems*, San Francisco, CA: ICS Press.

Ostrom, E. (2005), *Understanding Institutional Diversity*, Princeton, NJ: Princeton University Press.

Ostrom, E. (2007), 'A diagnostic approach for going beyond panaceas', *Proceedings of the National Academy of Sciences*, **104** (39), 15181–87.

Ostrom, E. (2009), 'A general framework for analyzing sustainability of social-ecological systems', *Science*, **325** (5939), 419–22.

Ostrom, E. and P. Benjamin (1991), 'Design principles and the performance of farmer-managed irrigation systems in Nepal', paper presented at the Third International Workshop of the

FMIS Network, 'Performance Measurement in Farmer-Managed Irrigation Systems', Mendoza, Argentina, 12–15 November.

Ostrom, E., L. Schroeder and S. Wynne (1993), *Institutional Incentives and Sustainable Development: Infrastructure Policies in Perspective*, Boulder, CO: Westview Press.

Ostrom, E., M. Janssen and J. Anderies (2007), 'Going beyond panaceas', *Proceedings of the National Academy of Sciences*, **104** (39), 15176–78.

Pierson, P. (2003), 'Big, slow-moving, and invisible: macro social process and contemporary political science', in J. Mahoney and D. Rueschemeyer (eds), *Comparative Historical Analysis in the Social Sciences*, Cambridge, UK: Cambridge University Press, pp. 177–207.

Pierson, P. (2004), *Politics in Time*, Princeton, NJ: Princeton University Press.

Poteete, A., M. Janssen and E. Ostrom (2010), *Working Together: Collective Action, the Commons, and Multiple Methods in Practice*, Princeton, NJ: Princeton University Press.

Pradhan, N. (1987), 'A farmer to farmer exchange training for improved irrigation management organized by DIHM's irriga-tion management center', memo, Kathmandu, Nepal: Irrigation Management Project.

Pradhan, P. (1989a), *Increasing Agricultural Production in Nepal: Role of Low-Cost Irrigation Development through Farmer Participation*, Kathmandu, Nepal: IIMI.

Pradhan, P. (1989b), *Patterns of Irrigation Organization: A Comparative Study of 21 Farmer-Managed Irrigation Systems*, Colombo, Sri Lanka: IIMI.

Pradhan, P. (2010), *Eroding Social Capital through Incompatible Legal and Institutional Regimes: Experiences from Irrigation Systems in Nepal*, Kathmandu, Nepal: Farmer Managed Irrigation Systems Promotion Trust.

Ragin, C.C. (1987), *The Comparative Method: Moving beyond Qualitative and Quantitative Strategies*, Berkeley: University of California Press.

Ragin, C.C. and J. Sonnett (2005), 'Between complexity and parsimony: limited diversity, counterfactual cases, and com-parative analysis', in S. Kropp and M. Minkenberg (eds), *Vergleichen in der Politikwissenschaft*, Wiesbaden: VS Verlag für Sozialwissenschaften, pp. 180–97.

Rihoux, B. (2008), 'Case-oriented configurational research: qualitative comparative analysis (QCA), fuzzy sets, and related techniques', in J.M. Box-Steffensmeier, H.R. Brady and D. Collier (eds), *The Oxford Handbook of Political Methodology*, Oxford: Oxford University Press, pp. 722–36.

Rodrik, D. (2007), *One Economics, Many Recipes*, Princeton, NJ: Princeton University Press.

Sengupta, N. (1991), *Managing Common Property: Irrigation in India and the Philippines*, New Delhi: Sage Publications.

Shivakoti, G. (1992), 'Variations in intervention, variations in result: assisting FMIS in Nepal', *Irrigation Management Network Paper No. 11*, London: Overseas Development Institute, Irrigation Management Network.

Shivakoti, G. (2006), 'Policies, institutions and governance challenges of irrigation in the twenty-first century', paper presented at the 11th biennial conference on 'Survival of the Commons: Mounting Challenges and New Realities', International Association for the Study of Common Property (IASCP), Bali, Indonesia, 19–23 June 2006.

Shivakoti, G.P. and E. Ostrom (eds) (2002), *Improving Irrigation Governance and Management in Nepal*, Oakland, CA: ICS Press.

Shivakoti, G., K. Giri and E. Ostrom (1992), 'Farmer-managed irrigation systems in Nepal: the impact of interventions', Research Report Series No. 18, Kathmandu, Nepal: HMG Ministry of Agriculture/Winrock International.

Shivakoti, G., D. Vermillion, W.F. Lam, E. Ostrom, U. Pradhan and R. Yoder (eds) (2005), *Asian Irrigation Systems in Transition: Responding to the Challenges Ahead*, New Delhi: Sage Publications.

Shivakumar, S. (2005), *The Constitution of Development*, New York: Palgrave Macmillan.

Shrestha, M.M. (1991), 'Experiences from SINKALAMA', in N. Ansari and Prachanda Pradhan (eds), *Assistance to FMIS: Experiences from Nepal*, Kathmandu, Nepal: Ministry of Water Resources, Department of Irrigation.

Shrestha, S.P. (1988), *Helping a Farmers' Organization: An Experience with Chiregad Irrigation Project*, Kathmandu, Nepal: IIMI.

Siy, R.Y., Jr. (1982), *Community Resource Management: Lessons from the Zanjera*, Quezon City: University of the Philippines Press.

Sparling, E.W. (1990), 'Asymmetry of incentives and information:

the problem of watercourse maintenance', in R.K. Sampath and R.A. Young (eds), *Social, Economic and Institutional Issues in Third World Irrigation Management*, Boulder, CO: Westview Press, pp. 195–213.

Tang, S.Y. (1992), *Institutions and Collective Action: Self-Governance in Irrigation*, San Francisco, CA: ICS Press.

Tan-Kim-Yong, U. (1987), 'Problems and strategies in management of communal irrigation systems: the experience in joint decision-making by farmers and agencies', in *Public Intervention in Farmer-Managed Irrigation Systems*, Colombo, Sri Lanka: IIMI and WECS of the MOWR, GON, pp. 277–92.

Turral, H. (1995), *Recent Trends in Irrigation Management: Changing Directions for the Public Sector*, Report No. 5, London: Overseas Development Institute.

UNDP (United Nations Development Programme) (2006), *Human Development Report 2006*, New York: Palgrave Macmillan.

UNESCO (United Nations Educational, Scientific and Cultural Organization)–WWAP (World Water Assessment Programme) (2006), *Water: A Shared Responsibility*, New York: Berghahn Books.

Uphoff, N. (1986), *Improving International Irrigation Management with Farmer Participation*, Boulder, CO: Westview Press.

Uphoff, N., P. Ramamurthy and R. Steiner (1991), *Managing Irrigation: Analyzing and Improving the Performance of Bureaucracies*, New Delhi: Sage Publications.

Vermillion, D.L. (1997), *Impacts of Irrigation Management Transfer: A Review of the Evidence*, Research Report No. 11, Colombo, Sri Lanka: IIMI.

Vermillion, D.L. (2005), 'Irrigation sector reform in Asia: from "participation with patronage" to "empowerment with account-ability"', in G. Shivakoti, W.F. Lam, D. Vermillion, E. Ostrom, U. Pradhan and R. Yoder (eds), *Asian Irrigation Systems in Transition: Responding to the Challenges Ahead*, New Delhi: Sage Publications, pp. 409–36.

Wade, R. (1982), 'The system of administrative and political corruption: canal irrigation in South India', *Journal of Development Studies*, **18** (3), 287–328.

Wade, R. (1987), 'Managing water managers: deterring expropriation or equity as a control mechanism', in W.R. Jordan (ed.), *Water and Water Policy in World Food Supplies*, College Station: Texas A&M University Press, pp. 177–83.

Wade, R. and D. Seckler (1990), 'Priority issues in the management of irrigation systems', in R.K. Sampath and R.A. Young (eds), *Social, Economic, and Institutional Issues in Third World Irrigation Management*, Boulder, CO: Westview Press, pp. 13–30.

WECS/IIMI (Water and Energy Commission Secretariat, Nepal, and International Irrigation Management Institute) (1990), *Assistance to Farmer-Managed Irrigation Systems: Results, Lessons, and Recommendations from an Action-Research Project*, Colombo, Sri Lanka: IIMI.

Whitford, A.B. and B.Y. Clark (2007), 'Designing property rights for water: mediating market, government, and corporation failures', *Policy Sciences*, **40** (4), 335–51.

Yoder, R. (1986), 'The performance of farmer-managed irrigation systems in the hills of Nepal', PhD dissertation, Cornell University.

Yoder, R. (1991a), 'Assistance to farmer managed irrigation systems: experiences from WECS/IIMI/Ford action research project in Indrawati watershed basin', in N. Ansari and P. Pradhan (eds), *Assistance to Farmer-Managed Irrigation Systems: Experience from Nepal*, Kathmandu, Nepal: Ministry of Water Resources, Department of Irrigation, Planning, Design, and Research Division, pp. 51–70.

Yoder, R. (1991b), 'Peer training as a way to motivate institutional change in farmer-managed irrigation systems', in *Proceedings of the Workshop on Democracy and Governance*, Decentralization: Finance & Management Project Report, Burlington, VT: Associates in Rural Development, pp. 53–67.

Yoder, R. (1994), *Locally Managed Irrigation Systems*, Colombo, Sri Lanka: IIMI.

Yoder, R. and S.B. Upadhyaya (1987), 'Reconnaissance/inventory study of irrigation systems in the Indrawati basin of Nepal', in *Irrigation Management in Nepal: Research Papers from a National Seminar*, Bharatpur, Nepal: IAAS/IIMI/Winrock International, pp. 15–23.

Yudelman, M. (1985), 'The World Bank and agricultural development: an insider's view', *World Resources Institute Paper* No. 1, Washington, DC: World Resources Institute.

Index

action-research 5, 30, 48–9, 51, 57, 97, 118
active participation 28, 57, 82–3, 103–4, 106, 112, 115, 119, 121
agencies 2, 4–5, 8, 10, 21–6, 29, 31–2, 34–9, 41–5, 47, 60, 63, 112, 114, 116, 119, 124–5
  credit investment 112
  donor 2, 8, 23, 29, 31
    external aid 3–4, 94, 117, 120, 121, 125
    external sources 107–8, 111
  governmental 10, 114
agricultural technologies 112
Andhi Khola 25
annual general meetings 99
Asia 1, 4–6, 8, 10, 13, 115, 125, 127–8, 135
Asian Development Bank 24, 103
assistance 3, 5–8, 10, 12–14, 16–19, 21, 23–31, 33–5, 37, 39, 41, 43, 45, 47–8, 50–54, 56–60, 62–4, 78–9, 86, 90, 92–4, 98, 101–3, 107–8, 112, 114–16, 118, 125, 128, 130, 134, 136
  external assistance 6, 17, 34, 79, 92–3, 98, 125, 130
  multifunctional assistance 112

best practices 3, 14, 65–6, 114, 119, 125
  trap 18–19, 94, 120
blueprints 6, 17, 57–8, 114, 118

canal 21, 34–5, 42, 44, 55, 70, 80–81, 83, 99, 102–6, 108–11
  contour canals 51

capacity 4, 6, 9, 21, 30, 35, 47, 54, 58, 112, 121, 123, 125–6
cash crop 111
collective action 3, 11–12, 16, 39–40, 43, 46, 68, 73, 79–84, 86, 89–90, 92, 95, 110, 117, 121, 126–7, 131–3, 135
  basis 46, 117
  problems 11–12
collective decision-making 112
command area 28, 45, 51–2, 59, 102, 111, 118
committee 26, 52, 79, 81, 98, 102, 106, 108, 122
Communist Party of Nepal (Maoist) 99
competition 85, 115–16, 122–4
complex
  dynamics of institutional change 16, 77
  factors 14, 75, 110
  interactions 7, 15
  processes 66, 119
  systems 7, 46, 125
complexity of social-ecological systems 7–8, 9, 15–16, 78–84, 93
control mechanisms 1–2
cooperation rules 98
coping
  with climate 2, 49, 51, 93–4
  with collective action 11, 12
  with complexity and change 78–84, 124
  with dual authority 97, 101
  with dynamic situations 112
  with physical environment 15, 94–5
  strategy 101